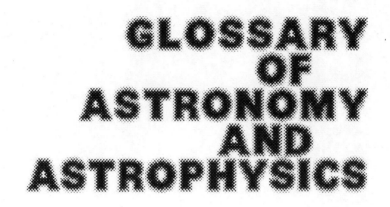

GLOSSARY
OF
ASTRONOMY
AND
ASTROPHYSICS

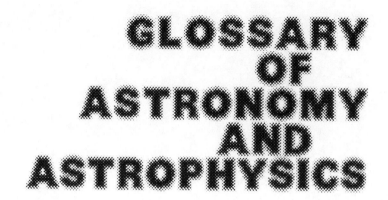

GLOSSARY OF ASTRONOMY AND ASTROPHYSICS

Jeanne Hopkins

Foreword by S. Chandrasekhar

THE UNIVERSITY OF CHICAGO PRESS
Chicago and London

This volume is published under the
auspices of *Astrophysical Journal*.

THE UNIVERSITY OF CHICAGO PRESS, CHICAGO 60637
THE UNIVERSITY OF CHICAGO PRESS, LTD., LONDON

© 1976 by The University of Chicago
All rights reserved. Published 1976
Printed in the United States of America

International Standard Book Number: 0-226-35172-6
Library of Congress Catalog Card Number: 75-14799

Foreword

The task of a copy editor of a scientific journal is in many ways an impossible one: to contend with the "illiteracies" (in the technical sense of Fowler) of an author without encroaching on his omniscience of scientific jargon is to walk a tightrope, requiring all the skill of an acrobat. Miss Jeanne Hopkins, who has edited the manuscripts for the *Astrophysical Journal* for many years now, has attempted to lighten her task by a genuine and an earnest desire to understand the meanings of the terms which the authors use while comprehending their content. With these latter objectives in view, she started compiling a glossary of the most commonly used astronomical and astrophysical terms, often transcribing the definitions given by different authors and asking me to referee (while I was still the Managing Editor) to resolve the conflicts which often arose. The glossary she started compiling in this fashion soon grew, and authors and editors to whom she showed the glossary were enthusiastic and recommended its publication. The present glossary arose in this way.

Clearly to compile a glossary that will stand the scrutiny of a tyro will require an astronomer of considerable scholarship; and it is quite unlikely that one with the requisite wide and detailed knowledge would want to take the time and effort. But Miss Hopkins, with the encouragement of several author-friends, has compiled a glossary with effectively the assistance of all the distinguished astronomers of this country through their papers in the *Astrophysical Journal*. I am sure that errors which may have crept into the glossary could be traced to one or other author in the *Astrophysical Journal!* But I am equally sure that when such an error is pointed out to Miss Hopkins, her response would be the same as Dr. Johnson's (when asked by a lady why he defined "pastern" as the "knee" of a horse, in his dictionary): "Ignorance, madam, pure ignorance."

Miss Hopkins deserves the thanks of all astronomers for her noble efforts in compiling this useful and valuable glossary.

S. Chandrasekhar

Acknowledgment

When I first started copyediting the *Astrophysical Journal* eight years ago, I felt like the girl in *Born Yesterday*. Many of the words either were totally unfamiliar or had a meaning different from any given in standard dictionaries. From sheer necessity I started compiling a list. Whenever I came across a felicitous explanation in a manuscript I was editing, I copied it down, frequently—and I now regret it—without citing the source. This compilation has become my vade mecum at work, and I hope it will prove a convenient quick reference for others. I would appreciate being informed of errors which have undoubtedly crept in.

I am grateful to the many *Astrophysical Journal* authors whose words I have quoted or paraphrased, and especially to the astronomers at the University of Chicago, who have patiently and graciously explained some abstruse term when I called them. I am deeply indebted to Dr. Dimitri Mihalas, who critically revised an earlier version of this glossary, and also to Professors David Van Blerkom and Thomas Adams for checking a later version and to Professor David Schramm for checking and rewriting many of the nucleosynthesis entries.

Most of all I want to express my deepest appreciation to Professor S. Chandrasekhar for initially communicating to me some of the fascination and excitement of astronomy, and for the many hours he spent going over the theoretical entries in the final version.

Jeanne Hopkins

A

A band   See Fraunhofer lines.

A-coefficient   See Einstein coefficient.

A-number   See atomic mass number.

A star   A star of spectral type A with a surface temperature of about 10,000 K, in whose spectrum the Balmer lines of hydrogen attain their greatest strength. Helium lines can no longer be seen. Some metallic lines are present; in late A stars the H and K lines of ionized calcium appear. A0 stars have a color index of zero. Examples of A stars are Vega and Sirius.

***ab* variables**   See Bailey types.

Am stars   Peculiar stars whose metallic lines are as strong as those of the F stars but whose hydrogen lines are so strong as to require that they be classed with the A stars. They are generally short-period ($<300^d$) spectroscopic binaries with high atmospheric turbulence and variable spectra, and are slower rotators than normal A stars.

Ap stars   Peculiar A-type stars ("magnetic" A stars) that show abnormally strong lines, sometimes of varying intensity, of certain ionized metals. Recent evidence indicates that all Ap stars are slow rotators compared with normal A stars.

aberration (of starlight)   Angular displacement in the apparent direction of a star, due to Earth's orbital motion (v_E) and the finite speed of light (c). The displacement $\Theta = \text{arc tan}\,(v_E/c) \approx 20''.49$. Thus to an Earth-based observer a star describes an ellipse on the celestial sphere with a semimajor axis of $20''.49$. (The eccentricity of the ellipse is zero—i.e., a circle—for a star on the ecliptic pole; for a star on the ecliptic plane the ellipse degenerates into a straight line.)

aberration (of an optical system)   Failure of a lens or mirror to achieve exact point-to-point correspondence between the object and the image.

ablation   Erosion of an object (generally a meteorite) by the friction generated when it passes through the Earth's atmosphere.

absolute magnitude (*M*)   Apparent magnitude that a star would have at a standard distance of 10 parsecs without absorption. The absolute magnitude of the Sun is $+4.85$. The absolute magnitude (g) of a solar-system body such as an asteroid is defined as the brightness at zero phase angle when the object is 1 AU from the Sun and 1 AU from the observer.

absolute zero   The point ($-273.16°$ C; 0 K) at which all thermal motion ceases and no heat is radiated.

absorption   Decrease in the intensity of radiation, representing energy converted into excitation or ionization of electrons in the region through which the radiation travels. As contrasted with monochromatic scattering (in which reemission occurs in all directions at the same frequency), the inverse process of emission refers to radiation that is reemitted in general in all directions and at all frequencies.

absorption coefficient (κ or k in cm^{-1})   Fraction of the incident radiation absorbed at a certain wavelength per unit thickness of the absorber. The absorption coefficient is in general a function of temperature, density, and chemical composition.

absorption edges   Sudden rises superposed on the smooth decrease of the curve of the attenuation coefficient, which cause the curve to have a typical sawtooth aspect. They generally occur at the limit of spectral lines.

absorption spectrum   Dark lines superposed on a continuous spectrum, caused by the absorption of light passing through a gas of lower temperature than the continuum light source.

accretion   A process by which a star accumulates matter as it moves through a dense cloud of interstellar gas; or, more generally, whereby matter surrounding a star flows toward it (as in close binaries).

Achernar (α Eridani)   A subgiant of spectral type B5, about 35 pc distant.

Achilles   Asteroid No. 588, a Trojan 60° ahead of Jupiter ($P = 11.98$ yr, $a = 5.2$ AU, $e = 0.15$, $i = 10°.3$). It was the first Trojan to be discovered (in 1906).

achromatic objective   A lens of two or more components with different refraction indices (e.g., crown glass and flint glass), used to correct for chromatic aberration.

active galaxy   See violent galaxy. Active galactic nuclei are very luminous (10^{43}–10^{46} ergs s^{-1}). Their energy output is in two forms: nonthermal continuum and thermal emission line.

active Sun   The Sun during its 11-year cycle of activity when spots, flares, prominences, and variations in radiofrequency radiation are at a maximum.

adiabatic index (γ)   The ratio of the fractional change in pressure to the fractional change in density as an element of fluid expands (or contracts) without exchange of heat with its surroundings.

advance of the perihelion   The slow rotation of the major axis of a planet's orbit in the same direction as the revolution of the planet itself, due to gravitational interactions with other planets and/or other effects (such as those due to general relativity).

aerolite (lit. "air stone.")   A stony meteorite, composed primarily of silicates. About 93 percent of all known falls are aerolites. They include the carbonaceous chondrites, other chondrites, and achondrites.

airglow (also called nightglow)   Light in the nighttime sky caused by the collision of atoms and molecules (primarily oxygen, OH, and Ne) in Earth's geocorona with charged particles and X-rays from the Sun or outer space. The airglow varies with time of night, latitude, and season. It is a minimum at zenith and maximum about 10° above the horizon.

air shower   A proliferation of secondary cosmic rays initiated either by primary cosmic rays or by high-enery (10^{14} eV) gamma rays.

albedo   Ratio of the total flux reflected in all directions to the total incident flux. (See Bond albedo; geometric albedo.)

Alcyone (η Tau)   The brightest star in the Pleiades (spectral type B5).

Aldebaran (α Tau)   A K5 III subgiant (a foreground star in the Hyades), about 21 parsecs distant. It has a faint M2 V companion. It is now known to be slowly and irregularly variable.

Alfvén speed (or Alfvén velocity) (v_A)   The speed at which hydromagnetic waves are propagated along a magnetic field: $(v_A) = B/(4\pi\rho)^{1/2}$.

Alfvén waves   Waves moving perpendicularly through a magnetic field. They are caused by the oscillation of magnetic lines of force by the motions of the fluid element around its equilibrium position, which in turn is caused by the interactions between density fluctuations and magnetic variations.

Algol (β Per)   An eclipsing system of at least three components (B8 V, K0, Am), about 25 pc distant. Period of components A and B is about 68.8 hours; period of components A, B, and C is about 1.9 years. Long term observations also indicate a massive, unseen fourth component with a period of about 190 years. Algol is also an erratic radio source of about 0.5 AU diameter.

aliasing   In a discrete Fourier transform, the overlapping of replicas of the basic transform, usually due to undersampling.

α-particle   The nucleus of a ^{4}He atom, consisting of two protons and two neutrons. Mass of α-particle 4.00260 amu.

α-particle nuclei   Nuclei formed by the α-process (q.v.)(see even-even nuclei).

α-process   A hypothetical process of nucleosynthesis (now considered obsolete terminology), which consisted of redistributing α-particles in the region from ^{20}Ne to ^{56}Fe (and perhaps slightly higher). The α-process has been replaced by explosive and nonexplosive C, O, and Si burning occurring in rapidly evolving or even explosive stages of stellar evolution which at higher temperatures and densities becomes the e-process (q.v.).

Altair (α Aql)   A bright ($m_v = 0.78$) A7 V star about 4.8 pc distant.

amagat   A unit of molar volume at 0° C and a pressure of 1 atmosphere. This unit varies slightly from one gas to another, but in general it corresponds to 2.24×10^4 cm^3. Also, a unit of density equal to 0.0446 gram mole per liter at 1 atm pressure.

Amalthea   Jupiter V, the innermost satellite of Jupiter. Diameter about 140 km; $i = 0°4$, $e = 0.0028$, period 0.498 days. Discovered by Barnard in 1892.

ampere   The SI unit of electric current. "The ampere is that constant current which, if maintained in two straight parallel conductors of infinite length, of negligible circular cross-section, and placed 1 meter apart in vacuum, would produce between these conductors a force equal to 2×10^{-7} newton per meter of length" (CIPM [1946], Resolution 2, approved by the 9th CGPM 1948). A current of 1 A is equivalent to the passage along the filament of a light bulb of about 6×10^{18} electronic charges per second.

amplitude (of wave motion)   Maximum displacement from the equilibrium position.

Andromeda Galaxy (M31, NGC 224)   A spiral galaxy (Sb in Hubble's classification; kS5 in Morgan's classification) in the Local Group, about 650–700 kpc distant ($M_V = -21$), visible to the naked eye as a fuzzy patch in the constellation of Andromeda. Total mass about 3.1×10^{11} $M\odot$; $i = 77°$, heliocentric velocity -180 km s^{-1}. Its nucleus exhibits noncircular gas motions. It is similar to but slightly larger than our Galaxy.

Andromeda I, II, III   Three dwarf spheroidal galaxies, in the Andromeda subgroup of the Local Group, discovered by van den Bergh in 1972. They are the intrinsically faintest members of the Local Group.

S Andromedae   A supernova seen in 1885 in the Andromeda Galaxy.

angstrom unit (Å)   A unit of length equal to 10^{-8} cm used in measuring wavelengths. 1 Å is about the diameter of a hydrogen atom (the diameter of a human hair is roughly 500,000 Å).

angular momentum (l or L)   The angular momentum of a system about a specified origin is the sum over all the particles in the system (or an integral over the different elements of the system if it is continuous) of the vector products of the radius vector joining each particle to the origin and the momentum of the particle. For a closed system it is conserved by virtue of the isotropy of space.

annihilation   See pair annihilation.

annular eclipse   An eclipse of the Sun in which the Moon is too far from Earth to block out the Sun completely, so that a ring of sunlight appears around the Moon.

anomalistic month   The interval (27.555 days) between two successive perigee passages of the Moon.

anomalistic year   The interval (365.2596 ephemeris days) between two successive perihelion passages of Earth.

anomalous Zeeman effect   Splitting of spectral lines into several components, in contrast to the normal Zeeman effect which results in only two distinct components. The anomalous Zeeman effect is due to the fact that the electrons in the magnetic field have opposite directions of spin.

anomaly   An angular value used to describe the position of one member of a binary system with respect to the other. The true anomaly of a star is the angular distance (as measured from the central body and in the direction of the star's motion) between periastron and the observed position of the star. The mean anomaly is the angular distance (measured in the same manner) between periastron and a fictitious body in the direction of the star, which is moving in a circular orbit with a period equal to that of the star.

ansae   The "handles," or extremities, of Saturn's rings as viewed from Earth; also, the extremities of a lenticular galaxy.

antalgol star   An old name for an RR Lyrae star.

antapex   The direction in the sky (in Columba) away from which the Sun seems to be moving (at a speed of 19.4 km s^{-1}) relative to general field stars in the Galaxy.

Antares (α Sco)   A red M1 Ib supergiant, about 125 parsecs distant. It has a B3 V companion, which is a radio source.

antenna gain   A measure of the directivity of a radio telescope. It is the ratio of the amount of power received in the direction the dish is pointing to the smaller amount of power from other directions in the sidelobes.

antenna temperature   A term used to describe the strength of a signal received from a radio source. It is the convolution of the true brightness distribution and the effective area of the antenna.

anticenter   The direction of the sky (in Auriga) opposite to that toward the center of the Milky Way.

anticoincidence counter   A particle counter in which the circuit has been designed so as not to register the passage of an ionizing particle through more than one counting tube.

antimatter   See antiparticle.

antiparticle   The absence of a particle in an otherwise infinite sea of negative energy.

apastron   The point in the orbit of one component of a binary system where it is farthest from the other.

aperture   The effective diameter of the primary mirror or lens of a telescope.

aperture efficiency (η_A)   The ratio of the effective aperture of the antenna, A, to its geometric aperture, $A_g = \pi d^2/4$. The beam and aperture efficiencies are related by $\eta_A = \eta_B \lambda^2/A_g\Omega_M$, where Ω_M is the solid angle of the main beam.

aperture function   In radio astronomy, a distribution of direction assignments applying to a uniform background.

aperture ratio (f)   The ratio of the aperture of a telescope to the focal length.

aperture synthesis   The spreading of parts of an antenna system over a pattern at several different distances while preserving the total collecting area of the system, in order to achieve better resolution. The technique employs one arm of a Mills cross and a small movable antenna that successively occupies the positions of the elements of the second arm of the cross.

apex   See solar apex.

aphelion   In the orbit of a solar system body, the most distant point from the Sun.

aplanatic system   A system of three lenses which, taken together, correct for spherical aberration, chromatic aberration, and coma.

apocenter   The point in the orbit of one component of a binary system which is farthest from the center of mass of the system.

apodization   A mathematical process performed on the data received from an interferometer before carrying out the calculations of the Fourier transformation to obtain the spectrum, in order to modify the instrumental response function.

Apollo asteroid   One of a small group of asteroids whose orbits intersect that of Earth. They are named for the prototype, Apollo ($P = 622^d$, $a = 1.486$ AU, $e = 0.57$, $i = 6°.4$).

apparent magnitude (m_v)   Measure of the observed brightness of a celestial object as seen from the Earth. It is a function of the star's intrinsic brightness, its distance from the observer, and the amount of absorption by interstellar matter between the star and the observer. The m_v of Sun, -26.5. A sixth-magnitude star is just barely visible to the naked eye.

apparent solar day   Interval between two successive culminations of the Sun—i.e., the period from apparent noon to apparent noon. The apparent solar day is longest in late December.

apparition   The period during which a celestial body is visible.

Appleton layers   See F layers.

appulse   A penumbral eclipse of the Moon.

apsidal motion   Rotation of the line of apsides (q.v.) in the plane of the orbit; (in a binary) precession of the line of apsides due to mutual tidal distortion.

apsides, line of   The major axis of an elliptical orbit.

arc spectra   The spectra of neutral atoms produced in a laboratory arc (cf. spark spectra).

Arcturus (α Boo)   An old subgiant disk star (K2 IIIp, $m_v = 0.06$) about 11 pc distant.

areas, law of   See Kepler's second law.

Argelander method (also called step method)   A method of classifying stars according to image size. If the sequence stars are labeled *a, b*, etc., in order of image size and if the image size of a variable appears to be, say, 0.7 of the way from sequence star *a* to sequence star *b*, its brightness is listed as *a7b*.

argument of the perihelion (ω)   Angular distance (measured in the plane of the object's orbit and in the direction of its motion) from the ascending node to the perihelion point.

Ariel   Second satellite of Uranus about 1600 km in diameter, discovered by Lassell in 1851. Period 2.52 days.

arm population   Young stars typical of those found in spiral arms (Population I stars).

array   In radio astronomy, an arrangement of antenna elements designed to produce a particular antenna pattern.

arrival time   See dispersion.

ascending node   In the orbit of a solar-system body, the point where the body crosses the ecliptic from south to north; for a star, out of the plane of the sky *toward* the observer.

ashen light   A faint glow from the unlit side of Venus when it is in the crescent phase. Its cause is unknown; it may be the Venusian analog to terrestrial airglow.

aspect ratio   Ratio of the major axis (e.g., of a rocket) to the minor axis. (Of a fusion device) ratio of the plasma diameter to the major diameter of the torus.

association   A sparsely populated grouping (mass range 10^2–10^3 $M \odot$) of very young, massive stars lying along a spiral arm of the Milky Way, whose spectral types or motions in the sky indicate a common origin. The star density is insufficient for gravitation to hold the group together against shear by differential galactic rotation, but the stars have not yet had time to disperse completely. OB associations are composed of stars of spectral types O-B2; T associations have many young T Tauri stars. The internationally approved designation for associations is the name of the constellation followed by an arabic numeral—e.g., Perseus OB2.

asteroid (also called minor planet)   A small planet-like body of the solar system, $\langle e \rangle \sim 0.15$, $\langle i \rangle \sim 9°.7$. More than 1800 have been catalogued, and probably millions of smaller ones exist, but their total mass would probably be less than 3 percent that of the Moon. Their densities are poorly known (about 2.6 g cm^{-3}), but they suggest a composition similar to carbonaceous chondrite. The bright asteroids are presumably original condensations and those fainter than about 14–15 mag are collision fragments. Asteroids and short-period comets have some orbital similarities.

asteroid belt   A region of space lying between Mars (1.5 AU) and Jupiter (5.2 AU), where the great majority of the asteroids are found. *None* of the belt asteroids have retrograde motion.

astigmatism   An aberration in optical systems produced when the object is far off axis (farther than in coma) and pairs of light rays in a single object plane do not focus in the same image plane.

astration   The processing of matter through stars.

astrometric binaries   See binary system.

astrometry   Measurement of the precise positions and motions of stars.

astronomical twilight   The period from sunset to the time that the Sun is 18° below the horizon; or the corresponding period before sunrise.

astronomical unit (AU)   The mean distance between the Earth and the Sun. The astronomical unit is defined as the length of the radius of the unperturbed circular orbit of a body of negligible mass moving around the Sun with a sidereal angular velocity of 0.017202098950 radian per day of 86,400 ephemeris seconds. 1

AU = 1.496 × 10^{13} cm ≈ 500 lt-sec.

atmosphere   Unit of pressure. 1 atm = 1.013 bars.

atmosphere (solar)   The gaseous outer layers of the Sun, including, from the deeper layers outward, the photosphere, the chromosphere, and the corona. The atmosphere constitutes those layers of the Sun that can be observed directly.

atmospheric extinction   Decrease in the intensity of light from a celestial body due to absorption and scattering by the Earth's atmosphere. The extinction increases from the zenith to the horizon and affects short wavelengths more than long wavelengths, so that objects near the horizon appear redder than they are at the zenith.

atomic mass number (A-number)   The number of protons plus neutrons in the nucleus of an atom.

atomic mass unit (amu)   In the notation of physics, adopted by international agreement in 1961, one-twelfth the mean mass of an atom of ^{12}C (including the orbital electrons). Because of the mass defect (q.v.), the amu is slightly less than the mass of a hydrogen atom, so one H atom has 1.007825 amu. 1 amu = 1.66 × 10^{-24} g. The energy equivalent of 1 amu is 931 MeV.

atomic number (Z-number) (also called charge number)   The number of protons in an atomic nucleus.

atomic time   Time based on the atomic second (see second). Atomic time was officially adopted 1972 January 1. From 1972 January 1 to 1974 January 1, 3 leap seconds had to be introduced to keep atomic time within 0.7 seconds of Universal Time.

atomic weight   The mean atomic mass of a particular element in atomic mass units.

attenuation   The falling off of the energy density of radiation with distance from the source, or with passage through an absorbing or scattering medium.

attenuation factor   In a rocket-borne or satellite-borne telescope, the ratio of the expected counting rate to the observed counting rate.

attitude   Position of a rocket with respect to the horizon or some other fixed reference plane.

atto-   A prefix meaning 10^{-18}.

AU   See astronomical unit.

Auger effect   A radiationless quantum jump that occurs in the X-ray region. When a K-electron is removed from an atom and an L-electron drops into the vacancy in the K-shell, the energy released in the latter transition goes not into radiation, but into the liberation of one of the remaining L-electrons.

AE Aurigae   An O9.5 V runaway star (q.v.).

α **Aurigae**   See Capella.

ε **Aurigae**   An eclipsing binary with an invisible supergiant companion. The primary is an extremely luminous A8 Ia supergiant of
30 $M \odot$ in a post-main-sequence stage of evolution; the secondary may be a collapsed star or black hole. The period of the
system is about 27 years. Probably on the order of 1 kpc distant.
It has at least six components.

RW Aurigae   A dG5e T Tauri star with a strong ultraviolet excess.

ζ **Aurigae stars**   In general, binaries with a K supergiant primary and
a main-sequence secondary.

aurora   Light radiated by ions in the Earth's atmosphere, mainly near
the geomagnetic poles, stimulated by bombardment by energetic
particles ejected from the Sun (see solar wind). Aurorae appear
about 2 days after a solar flare and reach their peak about 2
years after sunspot maximum.

autoionization (also called preionization)   A phenomenon occurring
when a discrete double-excitation state of an atom lies in the
ground-state continuum. In the autoionization process one of
the excited electrons is ejected, leaving the ion in an excited state
(see dielectronic recombination; see also Auger effect).

average life   See mean life.

Avogadro's number (6.02×10^{23})   The number of atoms in 12 grams
of ^{12}C; by extension, the number of atoms in a gram-atom (or
the number of molecules in a mole) of any substance.

azimuth   Angular distance from the north point eastward to the intersection of the celestial horizon with the vertical circle passing
through the object and the zenith.

azimuthal quantum number (k)   A measure of the minor axis of an
elliptic orbital of an electron according to the Bohr-Sommerfeld
theory.

B

B band   See Fraunhofer lines.

B-**coefficient**   See Einstein coefficient.

B galaxy   In Morgan's classification, a barred spiral.

b-**lines**   A triplet of spectral lines of neutral magnesium λλ5167–5184.

B star   Stars of spectral type B are blue-white stars with surface temperatures of about 11,000–28,000 K, whose spectra are characterized by absorption lines of neutral helium which reach their maximum intensity at B2. The Balmer lines of hydrogen are strong, and lines of singly ionized oxygen and other gases are also present. Examples are Rigel and Spica.

Ba II stars (also called barium stars)   Peculiar low-velocity, strong lined red-giant stars of spectral types G, K, and M, with abnormally large abundances of heavy s-process (but not r-process) elements. They are usually regarded as old disk stars of $\sim$1–2 $M\odot$.

background count   Unwanted counts due to background noise that must be subtracted from an observed number of counts in an experiment where atomic or nuclear particles coming from a source are being enumerated.

background noise   All the interference effects in a system which is producing, measuring, or recording a signal. Natural background noises arise from (a) galactic noise (synchrotron radiation), (b) thermal noise (receiver and isotropic background noise), (c) quantum noise (spontaneous emission or shot noise), and (d) star noise.

backscatter   Scattering of radiation (or particles) through angles greater than 90° with respect to the original direction of motion.

back warming   Heating of deeper layers due to overlying opacity.

Bailey types   A classification of RR Lyrae stars according to the shape and amplitude of their light variation (a, b, and c, although today types a and b are usually combined). The c-type stars have the smallest amplitude. (RRa: sharp rise to maximum; slow fall to minimum. RRc: Rise and fall equally long.)

Baily's beads   Small "beads" of sunlight (the "diamond ring" effect) which shine through the valleys on the limb of the Moon in the instant before (or after) totality in a solar eclipse. Named after the English astronomer Francis Baily who first observed them in 1836.

Baldet-Johnson bands   Spectral bands of the CO^+ radical.

Ballik-Ramsay bands   Spectral bands of the C_2 radical in the near infrared (0–0 at 1.7625 μ).

Balmer formula   A formula which represents the wavelengths of the various spectral series of hydrogen: $\lambda^{-1} = R(m^{-2} - n^{-2})$. The Balmer series is obtained by putting m equal to 2; the Lyman series by putting m equal to 1 (see Rydberg formula).

Balmer jump (also called Balmer discontinuity)   The sudden decrease in the intensity of the continuous spectrum at the limit of the

Balmer series of hydrogen at 3646 Å, representing the energy absorbed when electrons originally in the second energy level are ionized.

Balmer series   The spectral series associated with the second energy level of the hydrogen atom. The series lies in the visible portion of the spectrum. The transition from the third level to the second level yields the red Hα emission line at 6563 Å; Hβ is at 4861 Å; Hγ, at 4342 Å; Hδ, at 4101 Å. (Deuterium Hα is 1.785 Å shortward of hydrogen Hα.) He II Hα is at 1640 Å.

Bamberga   Asteroid 324 ($a = 2.80$ AU, $e = 0.36$, $i = 11°.2$). It is among the darkest known surfaces in the solar system. It is the only minor planet known to have an albedo less than 5 percent, and some astronomers think it may be larger than Pallas. Mean opposition magnitude $+11.41$, absolute magnitude $+8.14$. Rotation period 8^h(?). Meteorite class: carbonaceous chondrite.

band (molecular)   A series of closely spaced, often unresolved, emission or absorption lines found in the spectra of molecules. Each line represents an increment of energy due to a change in the rotational state of the molecule.

band head   The conspicuous sharp boundary which usually occurs at the head of a molecular band and which fades gradually toward either longer or shorter wavelengths, depending on the quadratic relation between frequency and rotational quantum number.

bandpass filter   A device used in radio astronomy for suppressing signals of unwanted frequencies without appreciably affecting the desired frequencies.

bandwidth   The width of the portion of the electromagnetic spectrum (the range of frequencies) that is permitted to pass through an electronic device (measured in cycles per second).

bar   The absolute cgs unit of pressure equal to 10^6 dyn cm^{-2}.

barium stars   See Ba II stars.

barn   A unit of area equal to 10^{-24} cm^2 used in measuring cross sections.

Barnard's loop   A huge nebular shell around the central portion of Orion.

Barnard's satellite   See Amalthea.

Barnard's star (BD+4°3561)   A faint M5 V optical binary (period about 25 years) about 1.83 pc distant ($\pi = 0°.548$) in the constellation of Ophiuchus. It has the largest proper motion known ($10°.25$ per annum). Long-term observations of its light curve suggest a possible third component with a mass about 1.2 that of Jupiter, although this observation has been challenged.

barometric law   The density distribution of gas in a plane-parallel, isothermal layer acted on by a uniform gravitational field: $\rho(z)$ = $\rho(0) \exp(-mg/kT)$.

barotropic gas   A gas in which the pressure is a function of the density only.

barred spiral galaxy (in Hubble's classification, SB; in Morgan's classification, B)   A spiral galaxy whose nucleus is in the shape of a bar, at the ends of which the spiral arms start. About one-fifth of spiral galaxies are barred spirals. First categorized by Hubble in 1936.

baryon   Heavy subatomic particle that interacts strongly in nuclei and that has a half-integral spin. Baryons obey the Fermi-Dirac statistics and include the nucleons and the so-called strange particles. Formerly, baryons heavier than the neutron were called hyperons; this term is seldom used today. All free baryons heavier than the proton are unstable and decay into end products, one of which is a proton.

Be stars (also called early-type emission stars)   Irregular variables of spectral type B (or occasionally O or A) with hydrogen emission lines in their spectra. The Be phenomenon involves rapid stellar rotation, circumstellar shells, and mass loss.

beam efficiency (of an antenna)   Fraction of the total received energy contained in the main beam (see aperture efficiency).

beat Cepheids   Dwarf Cepheids in which two or more almost identical periods exist which causes periodic amplitude fluctuations in their light curves. The "beat" period averages about 2 to 2 ½ hours.

Becklin-Neugebauer object (BN object)   An unresolved infrared point source (color temperature about 600 K) in the Orion Nebula. It is the brightest infrared object known at $\lambda \leq 10\ \mu$, and is not coincident with any distinctive optical or radio continuum feature. Probably a collapsing protostar of 5–10 $M \odot$. Discovered in 1966.

Beehive Cluster   See Praesepe.

Bellatrix (γ Orionis)   A B2 III star 80 pc distant.

Bernoulli probability   See binomial probability.

Bernoulli's theorem   Along a streamline the total energy per unit mass (including the internal energy and the pressure head p/ρ) of an element of fluid remains constant as it moves.

Bessel equation   A linear second-order differential equation, the solutions to which are expressible in mathematical functions known as Bessel functions.

β-decay Emission of an electron and an antineutrino (or a positron and a neutrino) by a radioactive nucleus by any one of several processes, e.g., the spontaneous β-decay of a free neutron ($n \rightarrow p + e^- + \bar{\nu}$). The A-number is unchanged, but the Z-number is increased (or decreased) by 1. Beta-decay is a so-called weak interaction (q.v.). Since electrons of all energies (up to a certain maximum) are emitted in β-decay, this process exhibits a continuous spectrum (unlike α-particle emission, which exhibits a line spectrum).

β-particle An electron or a positron emitted from an excited nucleus when it returns to its ground state via β-decay.

β-transition See spectral lines.

Betelgeuse (α Ori) A red semiregular variable supergiant (M2 Iab) about 500 pc distant. Betelgeuse is also a strong infrared emitter—at 2 μ the brightest in the sky.

Bethe-Weizsäcker cycle See carbon cycle.

BF$_3$ counter A proportional counter, filled with the gas boron trifluoride, designed to count neutrons.

Bianchi cosmology A cosmology which, unlike the Friedmann cosmology, dispenses with the notion of isotropy and considers homogeneous spaces. The different forms of homogeneous cosmologies that are possible are classified, according to the structure parameters of the associated groups, into nine (or 10 if a special case is included) classes.

big-bang model A model of the Universe which started with an initial singularity. The Friedmann model of a homogeneous, isotropic universe (composed of adiabatically expanding matter and radiation, as a result of a primeval explosion) is the standard example.

binary system A system of two stars orbiting around a common center of gravity. Visual binaries are those whose components can be resolved telescopically (i.e., angular separation >0.5) and which have detectable orbital motion. Astrometric binaries are those whose dual nature can be deduced from their variable proper motion; spectroscopic binaries, those whose dual nature can be deduced from their variable radial velocity. At least half of the stars in the solar neighborhood are members of binary (or multiple) systems. (See photometric binaries; optical pairs.)

binding energy The energy required to break up a system. In particular, the binding energy of an atomic nucleus is the energy released in the formation of the nucleus. The most strongly bound nuclei are those with atomic weights between about 50 and 65 (the iron group). Lighter nuclei are less strongly bound because of their larger surface-to-volume ratios; heavier nuclei, because the effects of Coulomb repulsion increase with the nuclear charge.

binomial probability (also called Bernoulli probability)   The probability that a particular result will be obtained in a given number of trials.

Birkhoff's theorem   Every centrally symmetric geometry which is free of mass-energy is static and identical up to a coordinate transformation with the geometry defined by the Schwarzschild metric.

bit   In computer terminology, a shortened form for binary digit (0 or 1).

Blaauw mechanism   A mechanism advanced to explain the disruption of a binary system by the decrease in the gravitational binding force when an ejected shell overtakes the secondary component.

blackbody   An idealized body which absorbs radiation of all wavelengths incident on it. (Because it is a perfect absorber, it is also a perfect emitter.) The radiation emitted by a blackbody is a function of temperature only.

blackbody radiation (sometimes called thermal radiation)   Radiation whose spectral intensity distribution is that of a blackbody in accordance with Planck's law.

black dwarf   The final stage in the evolution of a star of roughly 1 $M \odot$. It is a mass of cold, electron-degenerate gas, and can no longer radiate energy, because the whole star is in its lowest energy state. No black dwarfs have ever been observed. Also, an object ($M < 0.085 \, M \odot$) that is not massive enough to achieve nuclear chain reactions.

black hole   A gravitationally collapsed mass inside the Schwarzschild radius (q.v.), from which no light, matter, or signal of any kind can escape. A black hole occurs when the escape velocity of a body becomes the velocity of light ($2GM/R = c^2$). If an object with the mass of the Sun had a radius of 2.5 km, it would be a black hole. Black holes represent one of the possible endpoints of stellar evolution for stars very much more massive than the Chandrasekhar limit.

blazed grating   Diffraction grating so ruled that the reflected light is concentrated into only a few orders, or even a single order, of the spectrum.

blue edge (of the RR Lyrae instability strip)   The curve on the H-R diagram that is traced out by the maximum temperature at which a stellar model is unstable against small-amplitude pulsations as the luminosity is varied. The position of this borderline is a function of mass and composition. Iben defines the blue edge as that point where the growth rate for pulsation is zero.

blue halo stars   Hot stars that are in the horizontal-branch, post-horizontal-branch, and post-asymptotic branch phases of evolution.

blue haze   A condition in the Martian atmosphere which sometimes makes it opaque to radiation in the blue-violet end of the visible spectrum.

blue horizontal-branch stars   Population II stars (B3–A0) in the galactic halo, characterized by strong, sharp hydrogen lines and large Balmer jump, and very weak lines of all other elements (see also HZ stars).

blueshift   Shift of spectral lines toward shorter wavelengths in the spectrum, which occurs when the source is approaching.

blue stragglers   Stars (in a cluster) which fall close to the cluster's extrapolated main sequence but which occur a few magnitudes above its turnoff point.

BN object   See Becklin-Neugebauer object.

Bohr atom   The model of an atom whose electrons are pictured as describing "Keplerian" orbits about the central nucleus.

Bohr magneton (μ_0 or μ_B)   Magnetic moment of an electron in the first Bohr orbit ($\mu_0 = eh/4\pi mc$). It is a unit representing the minimum amount of magnetism which can be caused by the revolution of an electron around an atomic nucleus. 1 Bohr magneton $= 0.92 \times 10^{-20}$ ergs per gauss.

Bohr radius ($a_0 = \hbar^2/me^2$)   The radius of the orbit of the hydrogen electron in its ground state (0.528 Å). The electron makes 6.6×10^{15} revolutions s^{-1} (velocity of electron 2.19×10^8 cm s^{-1}).

Bok globule   A compact, spherical dark nebula. Bok globules have characteristic radii of 10^3–10^5 AU. Estimates of their mass suggest that their density is too low for gravitational collapse. They tend to lie in regions of much dust but less gas than would be expected for star-forming regions.

bolide   See meteor.

bolometric absolute magnitude (M_{bol})   A measure of the total amount of energy radiated by a star at all wavelengths. M_{bol} of Sun = 4.72 mag. The fraction of total energy emitted by a very blue or very red star that lies in the visible range may differ from the total energy by 4 or 5 mag—i.e., only a few percent of the energy lies in the visible.

bolometric correction (B.C.)   The visual (or photovisual) magnitude minus the bolometric magnitude of a star. It is always a positive number.

Boltzmann's constant (k)   The constant of proportionality relating the mean kinetic energy of an atom to its absolute temperature: $k = 1.38 \times 10^{-16}$ ergs per kelvin.

Boltzmann factor   The factor $e^{-E/kT}$ involved in the probability for atoms having an excitation energy E at temperature T.

Boltzmann-Saha theory   A theory describing the distribution of atoms of partially ionized matter over possible excitation and ionization states, in the limit of low density (cf. Thomas-Fermi theory).

Bond albedo   Fraction of the total incident light reflected by a spherical body. It is equal to the phase integral, multiplied by the ratio of its brightness at zero phase angle to the brightness it would have if it were a perfectly diffusing disk.

α Bootis   See Arcturus.

λ Bootis stars   A type of young (usually early A), weak-lined, metal-poor stars with low radial velocities.

Born approximation   An approach to collision problems by using perturbation methods (q.v.). In collisional excitation the Born approximation becomes valid when the incident energy is some 50 times larger than the excitation energy.

Born-Oppenheimer approximation   An approximation that treats the motion of an electron under the attraction of two free nuclei by regarding the nuclei (because of their greater mass and consequent smaller velocities) as fixed.

Bose-Einstein nuclei   Nuclei of even A-number (i.e., those with integral spin) (cf. Fermi-Dirac nuclei). Bose-Einstein nuclei do not obey the exclusion principle, and their ground state has zero angular momentum.

boson   A subatomic particle whose spin is an integral multiple of $\hbar$ (cf. fermion). Bosons include the photons, the pions, the gravitons, and all Bose-Einstein nuclei. Boson number is not conserved.

bound-bound transitions   Transitions between energy levels of an electron bound to a nucleus (the electron is bound both before and after the transition).

bound-free transitions   Transitions in which a bound electron in any energy level is liberated.

Boussinesq equations   Hydrodynamic equations often used to analyze the onset of convection in a fluid by allowing for the variations of density only insofar as buoyancy forces are concerned.

Bowen fluorescence mechanism   A mechanism first discovered by Bowen which explains the anomalously strong lines of O III in the spectra of some planetary nebulae as fluorescence involving the radiative excitation of the $2p3d$ $^3P^o_2$ level of O^{2+} (54.71 eV) from the $2p^2$ $^3P_2$ state in the ground term by He II Lyman-α photons (54.17 eV).

Boyle's law   The pressure of an ideal gas kept at constant temperature varies inversely as the volume, i.e., directly as the density.

Bp stars   Peculiar B stars whose spectra show a deficiency in helium.

Brackett series   The spectral series associated with the fourth energy level of the hydrogen atom. Bα is at 40512 Å. (He II Bα is at 10124 Å; see Pickering series).

Bragg angle   Glancing angle between an incident X-ray beam and a given set of crystal planes for which the secondary X-radiation from the planes combines to give a single reflected beam.

braking parameter (n) (of a pulsar)   A quantity describing the rate of slowdown of a pulsar, $dE/dt = A\omega^n$, where the braking index $n = \omega\dddot{\omega}/\omega^2$.

branching ratio   Ratio between the numbers of atoms starting from a given initial state which undergo two different types of transitions, perhaps, or between different bound states.

Brans-Dicke theory   A scalar-tensor modification of the general theory of relativity by the introduction of a scalar field (in Einstein's theory, gravitation is described by a single field quantity, a tensor).

Breit-Wigner equation   An equation relating the cross section in a nuclear reaction to the energy of the incident particle.

bremsstrahlung (lit., "deceleration radiation")   Radiation emitted or absorbed when a free electron is accelerated in the field of an atomic nucleus but remains in a hyperbolic orbit without being captured. Since bremsstrahlung is not quantized, photons of any wavelength can be emitted or absorbed. (Also called a free-free transition because the electron is free both before and after the transition.)

bright points   Bright regions (in X-ray and XUV) observed on the Sun during *Skylab* missions. They are distributed fairly uniformly over the disk, and have diameters of about 20,000 km, mean lifetimes of about 8 hours, and electron temperatures of about $1-2 \times 10^6$ K.

bright ring   See Saturn's rings.

brightness   Measure of the luminosity of a body in a given spectral region.

brightness distribution   A statistical distribution based on brightness, or the distribution of brightness over the surface of an object.

brightness temperature   The temperature that a blackbody would have to have to emit radiation of the observed intensity at a given wavelength. Brightness temperature in radio astronomy is equivalent to specific intensity in optical astronomy.

Brillouin scattering   Slight changes in the frequency of radiation, caused by reflection or scattering from the high-frequency sound waves that arise from thermal vibrations of atoms in the medium.

Brillouin zone   A continuous ensemble of all energies and wave functions which may be obtained from one atomic energy level in a metallic-crystal lattice.

burst (cosmic-ray)   A sudden intense ionization, apparently due to the effect of cosmic rays on matter, and often giving rise to great numbers of ion pairs at once.

burst (solar)   Suddenly enhanced nonthermal radio emission from the high solar corona immediately following a solar flare, probably due to energetic electrons trapped in the coronal magnetic field. Bursts are divided into several types, depending on their time frequency characteristics (type III is the most common). They are classified on a scale of importance ranging from -1 (least important) to $+3$. Bursts are generally attributed to a sudden acceleration of some 10^{35-36} electrons to energies greater than 100 keV in less than 1 second.

butterfly diagram   Plot of heliographic latitude of sunspots versus time, developed by Maunder in 1904 to illustrate the solar cycle.

Bw stars   B stars with weak helium lines—i.e., B stars which, if classified according to their colors, would have helium lines too weak for the classification, and which, if classified according to their helium lines, would have colors too blue for their spectral type.

C

C galaxies   In the Yerkes 1974 system, small, high surface-brightness galaxies which are slightly resolved on medium- and large-scale photographs.

C stars   A class of carbon stars (q.v.), defined by Morgan and Keenan to replace the Harvard R and N spectral classes.

C-S stars   Group characteristics are: strong bands of CN, outstandingly strong absorption near the Na D lines, usually sufficient structure in the 6400–6500 Å region to suggest ZrO.

calcium star   Old name for an F star.

California Nebula (IC 1499)   An H II region ionized by Zeta Persei.

Callisto   A Galilean satellite (J IV) of Jupiter, about 5050 km in diameter. Orbital and rotation period 16.7 days ($e = 0.0075$; $i = 0°.3$). It has the lowest density (1.7 g cm^{-3}), lowest albedo (0.15), and *highest* temperature (156 K) of any of the four.

Z Camelopardalis stars   A class of dwarf novae (q.v.) with standstills in their light curves. Z Cam itself is a semidetached binary (period 7^h21^m) consisting of a dG1 star and a hot white dwarf or a hot blue subdwarf which is probably degenerate. Mean time between eruptions, 20 days. Peak-to-peak amplitude, about 0.5 mag.

candela   The SI unit of luminous intensity, defined as "the luminous intensity, in the perpendicular direction, of a surface of 1/600,000 square meter of a blackbody at the temperature of freezing platinum under a pressure of 101,325 newtons per square meter." (13th CGPM [1967], Resolution 5.)

α Canis Majoris   See Sirius.

β Canis Majoris star   See Beta Cephei star.

VY Canis Majoris   A peculiar cM3e irregular variable with an extremely strong infrared excess, presumably due to a circumstellar dust shell. It is a class 2b OH emitter, and CO and H$_2$O have been identified in its spectrum. It is a multiple star with at least six components, surrounded by a small reflection nebula, about 1.5 kpc distant, in the galactic plane. It may be a pre-main-sequence star, or it may be a highly evolved object like an extremely young planetary nebula.

α Canis Minoris   See Procyon.

canonical change   A periodic change in one of the components of an orbit (cf. secular change).

Canopus (α Car)   A type F0 Ib supergiant, about 55 pc distant, the second brightest star in the southern sky.

α^2 CVn star   See spectrum variable. α^2 CVn has a period of 5.469 days. Its spectrum shows strong, profuse lines of rare earths, iron-peak elements, and Si.

AM Canum Venaticorum (HZ 29)   A peculiar blue variable. It may be an accreting, semidetached binary white dwarf system with a period of about 18 minutes (0.012 days).

Capella (α Aur)   A spectroscopic triple (F8–G0 III, G5 III, M5 V) with a period of 104.023 days, about 13 pc distant (1974 parallax

(0.̊079). It has a high lithium content and a nearly circular orbit. It may be an X-ray source.

carbon cycle (also called CN cycle or Bethe-Weizsäcker cycle) (discovered in 1938-39)   A series of nuclear reactions in which carbon is used as a catalyst to transform hydrogen into helium: $^{12}C(p,\gamma)^{13}N(p,\gamma)$ $^{14}O(\beta^+\nu)^{14}N(p,\gamma)^{15}O(\beta^+\nu)^{15}N(p,\alpha)^{12}C$. The carbon cycle can take place only if the necessary C and N nuclei are present, and it requires higher temperatures (15–20 million kelvins) and is far more temperature-dependent ($E \propto T^{15}$) than the proton-proton chain ($E \propto T^4$). The cycle yields 26.7 MeV of energy. (On the average, 1.7 MeV of this energy is carried away because of neutrino losses.)

CNO bi-cycle   Similar to the CN cycle, except that it also includes a cycle in which the next-to-last step becomes $^{15}N(p,\gamma)^{16}O(p,\gamma)$ $^{17}F(\beta^+\nu)^{17}O(p,\alpha)^{14}N$. This reaction occurs once in about 2,000 CN cycles. For main-sequence stars greater than a few solar masses, hydrogen burning by the CNO bi-cycle is the main source of energy. (It produces about 2% of the solar energy.)

CNO tri-cycle   Similar to the CNO bi-cycle, with the addition of the cycle $^{17}O(p,\gamma)^{18}F(\beta^+\nu)^{18}O(p,\alpha)^{15}N$.

carbon detonation supernova model   A supernova model involving the explosive ignition of carbon in the high-density ($10^8 - 10^{10}$ g cm^{-3}), electron-degenerate carbon-oxygen core of a $6\pm2 - 7\pm2$ $M\odot$ star by the formation and propagation of a detonation wave. A carbon-detonation supernova seems to leave no dense remnant and converts its C-O core entirely to iron.

carbon stars   In the HD system, a rather loose category of peculiar red-giant stars, usually of spectral types R and N, whose spectra show strong bands of C_2, CN, or other carbon compounds and unusually high abundances of lithium. Carbon stars resemble S stars in the relative proportion of heavy and light metals, but they contain so much carbon that these bands dominate their spectra (see also C stars). (C2,0. The number following the comma is an abundance parameter.)

carbonaceous chondrites   Chondrites (q.v.) characterized by the presence of carbon compounds. They are the most primitive samples of matter in the solar system.

Carina OB 2   A rich association of OB stars near η Carinae.

α Carinae   See Canopus.

η Carinae   A peculiar nova-like variable about 2 kpc distant. For 50 years in the middle of the nineteenth century it was the second brightest star in the southern sky, reaching magnitude -1 in

1843. Presently its visual apparent magnitude is $+7$ (although at 20 μ it is still the brightest source in the sky). It may be a "slow supernova" with its slowness due to the large size of the parent star.

Carter's theorem   Sequences of axisymmetric metrics external to black holes must be disjoint, i.e., have no members in common.

Cassegrain focus   An optical arrangement in which light rays striking the parabolic concave primary mirror of a reflecting telescope are reflected to the hyperbolic convex secondary mirror, and re-reflected through a hole bored in the primary to a focus behind it.

Cassini's division   A gap about 1800 km wide between the outermost rings of Saturn. It was discovered by Cassini in 1675. The period of a particle in Cassini's division is about two-thirds that of Janus, one-half that of Mimas, one-third that of Enceladus, and one-quarter that of Tethys.

Cassiopeia A (3C 461)   A radio source in Cassiopeia, the strongest extrasolar source in the sky, perhaps 3 kpc distant, believed to be the remnant of a Type II supernova whose light reached Earth about 1667. Optically it is a faint nebula. It has an expansion velocity of about 800 km s^{-1} and a mass of a few solar masses. It is also an extended source of soft X-rays (3U 2321+58).

AO Cassiopeiae   A binary in which the larger, less massive, hot primary is highly distorted, and in which rapid mass exchange is occurring.

B Cassiopeiae   See Tycho's star.

WZ Cassiopeiae   A carbon star (the most super-rich carbon star known) with a very high abundance of lithium. Its effective temperature is 2420 K.

Castor (α Geminorum)   A visual triple system about 14 pc distant. Each component is itself a spectroscopic binary. Component A is A1 V, with a period of 9.22 days; component B is Am5 with a period of 2.93 days. Period of components A and B is about 380 years. Component C (YY Gem), a flare star, is a double-lined eclipsing binary with a period of 0.814 days. Both components are dM1e, and both components exhibit flares.

cataclysmic variable   A type of variable including flare stars and novae (common, recurrent, and dwarf), all of which are believed to be very close binary systems in which hydrogen-rich matter flows from a late-type star onto a hot white-dwarf primary.

Cauchy dispersion formula   An approximate empirical formula for the index of refraction n as a function of wavelength: $n = A + B/\lambda^2 + C/\lambda^4 + \ldots$, where $A, B, C, \ldots$ are constants depending on the refracting medium.

cD galaxy   In Morgan's classification, a supergiant elliptical galaxy with a large, faint halo; an outstandingly large, luminous D galaxy. cD galaxies occur centrally located in rich clusters of galaxies.

celestial equator   The great circle where the plane of Earth's equator, if extended, would touch the celestial sphere.

celestial longitude   Angular distance along the ecliptic from the vernal equinox eastward.

celestial meridian   The great circle on the celestial sphere which passes through the celestial poles and the zenith of the observer.

celestial poles   The two points at which the Earth's axis of rotation, if extended, would intersect the celestial sphere.

α Centauri (Rigil Kent)   A binary system (G2 V, K5 V) 1.3 pc distant. Period of system about 80 years. Parallax $0''.754$, proper motion $3''.68$ per year.

Proxima Centauri   An eleventh magnitude ($M_{bol} = 11.66$ mag) star, probably associated with the α Cen system. It is a flare star of spectral type dM4e with a parallax of $0''.765$, which makes it our closest known stellar neighbor. $M = 0.1 \, M\odot$; $R = 1.3 \times 10^{10}$ cm.

ω Centauri   A metal-poor halo-population globular cluster of more than $3 \times 10^6 \, M\odot$, according to Poveda. It is the closest known globular cluster (about 5.2 kpc distant) and is barely visible to the naked eye in Earth's southern hemisphere.

Centaurus A   A strong radio source. Optically, it is an elliptical galaxy (NGC 5128) with a dark obscuring lane. It is the nearest known violent galaxy. Probably about 4 Mpc distant. It is also an X-ray source (3U 1322−42).

Centaurus cluster (3U 1247−41)   A cluster of galaxies about 200 Mpc distant. It is also an extended X-ray source. Its radio counterpart is compact and located inside NGC 4696.

Centaurus X-3 (3U 1118−60)   A pulsating (period 4.8 s) binary X-ray source in the galactic plane, recently found to be a member of an occulting binary system ($e < 0.002$, period of system 2.087 days; X-ray eclipse lasts 0.488 days). Optical component is Krzeminski's star, a B0 giant or supergiant, about 5–10 kpc distant. The X-ray component is probably a rotating neutron star of about 0.65–0.83 $M\odot$. Cen X-3 is speeding up at a rate of about 1 part in 10^3–10^5 per year and will at this rate fall into its companion in about 1000 years.

Cen X-2 and Cen X-4 are sporadic X-ray sources.

β Cephei stars (also called β Canis Majoris stars)   A small group of short-period ($P = 3\frac{1}{2}$ to 6 hr) pulsating variables (O9–B3) lying slightly above the upper main sequence. They have a doubly periodic light curve, and are confined within a narrow band of the H-R diagram which lies near the end of core hydrogen-burning stars of roughly 10–20 $M\odot$. Beta Cephei itself has at least three components.

VV Cephei stars   Eclipsing binaries with M supergiant primaries and blue (usually B) supergiant or giant secondaries. They have a rich emission spectrum. Sandage (1974) suggests $M_V = -7.3$ for the M2p component of VV Cep.

Cepheid variable   One of a group of very luminous supergiant pulsating stars named for the prototype δ Cep. Type I (or classical Cepheids) are extreme Population I with characteristic periods of 5–10 days. They are about 4 times more luminous ($<M_V> = -0.5$ to -6) than those of type II, probably because of their higher metal content (although mass may also be a factor). Type II (or W Virginis stars) are Population II ($<M_V> = 0$ to -3) with characteristic periods of 10–30 days, and are primarily found in globular clusters. The luminosities of all Cepheids are proportional to their periods, but a different P-L relation applies to each type. No Cepheid is near enough for an accurate trigonometric parallax (Polaris is the nearest). Cepheids are useful distance indicators to about 3 Mpc.

Cerenkov radiation   Visible (and more energetic) radiation caused by an electromagnetic shock wave arising from charged particles moving with velocities greater than the speed of light in the medium. (It is the electromagnetic analog to an acoustical shock wave, or sonic boom.) The particles will continue to lose energy by radiation until their velocity is less than the speed of light in the medium.

Ceres   Largest of the known asteroids, and the first to be discovered (by Piazzi in 1801). $R \approx 510$ km, mean distance from Sun 2.7673 AU, $e = 0.079$, $i = 10°.6$. Rotation period 0.38 days, sidereal period 1,682 days, synodic period 466.6 days. Photographic albedo 0.06. Mean orbital speed 17.9 km s^{-1}. Mass 1.17×10^{24} g. Spectrum suggests carbonaceous chondrite.

o Ceti   See Mira.

τ Ceti   A G8 Vp star about 3.6 pc distant.

UV Ceti stars   Late-type dwarfs (dKe–dMe) with spectra showing hydrogen emission lines. UV Cet itself is a faint M6e V flare star (component B of Luyten 726-8) of very low mass (0.15 $M\odot$), 2.8 pc distant. Like other flare stars, it is a member of a binary system in which both components are of nearly equal brightness (M_V = 15.3 and 15.8). Period of the system is about 26.5 years (angular separation $1''.0$, e = 0.615). Radio flares have also been observed.

Cetus Arc   A gaseous nebula, probably about 100 pc distant, centered on or near β Peg. It may be a supernova remnant.

Chandler period   The period of the variation of the celestial poles (about 416–433 days, with a peak at 428 days). Pole wandering (by as much as 15 meters from its mean position) causes minute variations in the meridian.

Chandrasekhar limit   A limiting mass for white dwarfs. If the mass exceeds this critical mass (1.44 solar masses, for the expected mean molecular weight of 2), the load of the overlying layers will be so great that degeneracy pressure will be unable to support it, and no configuration will be stable.

Chandrasekhar-Schönberg limit   The mass limit for an isothermal core. In order to maintain its luminosity by hydrogen burning just outside the isothermal core, the star must keep a high temperature and a high pressure at the surface of the core. When the helium core exceeds about 12% of the star's total mass, the star can no longer adjust by small changes, but must drastically increase in radius and move rapidly from the main sequence.

Chapman's equation   An equation expressing the velocity of a gas in terms of certain molecular constants.

Chapman-Jouguet detonation   A detonation in which the velocity of the shock front with respect to the material *behind* it is equal to the corresponding sound velocity.

characteristic value   See eigenvalue.

charf   A permanent blemish on an image-tube phosphor.

charge conjugation   The technical term for mathematical operations which interchange particles and antiparticles.

charge multiplet   A group of particles (such as the two nucleons or the three pions) which differ in electrical charge but which are nearly identical in mass and other respects (such as lifetime and angular momentum) and which seem to experience identical nuclear forces.

charge number   See atomic number.

Charles's law   The pressure of an ideal gas at constant volume varies directly as the absolute temperature.

χ^2 **test**   A least-squares statistical test that measures the probability of randomness in a distribution.

chondrite   A stony meteorite usually characterized by the presence of chondrules (q.v.). (Type I carbonaceous chondrites contain no chondrules.)

chondrules   Small spherical grains varying from microscopic size to the size of a pea, usually composed of iron, aluminum, or magnesium silicates. They occur in abundance in primitive stony meteorites. Chondrules show evidence that they were formed at about the same time as the planets—it has been suggested that they formed in the solar nebula by impact between high-velocity grains. True chondrules have never yet been observed in terrestrial rocks.

chromatic aberration   A defect of refracting telescopes whereby light of different colors is focused at different distances behind the objective. Blue light is refracted more than red light and hence comes to a focus inside that of red light. Images are then surrounded by a rainbow of colors.

chromosphere   The part of the solar atmosphere between the photosphere and the corona. It consists of two rather well defined zones: the lower chromosphere extends to about 4000 km ($\rho \approx 10^{-8}$ to 10^{-13}g cm^{-3}) and consists of cool ($\sim$7500 K) neutral hydrogen; the upper chromosphere extends to about 12,000 km ($\rho \approx 10^{-16}$ g cm^{-3}) and consists of hot (10^6 K), ionized hydrogen. It has an emission spectrum (see flash spectrum).

chromospheric network   A large-scale cellular pattern along the boundaries of which lie bright and dark mottles seen in Hα and other regions.

Circinus X-1 (3U 1516$-$56)   A highly variable X-ray source. Many of its properties are similar to those of Cygnus X-1.

cislunar   An adjective referring to the region of space between the Earth and the Moon.

Clapeyron's equation   A fundamental relation between the temperature at which an inter-phase transition occurs, the change in heat content, and the change in volume.

closed universe   A world model in which the expansion velocity of the original big bang was less than the "escape velocity" of the Universe. In this model the rate of expansion will steadily decrease and come to a halt, and then the Universe will start to contract. The critical density required to close the Universe is 5 $\times$ 10^{-30} g cm^{-3} (or about 3×10^{-6} H atoms cm^{-3}) if $H_0 = 55$ km s^{-1} Mpc^{-1}. The present mass density appears to be less than the critical density, but this is still an open question.

cluster of galaxies   An aggregate of galaxies. Bautz and Morgan divide them into three morphological types: type I contains a supergiant cD galaxy; type III contains no members significantly brighter than the general bright population. Coma is type II, Virgo is type III. Rood recognizes three types: A compact group (e.g., Stephan's Quartet) contains a few galaxies separated by a few galaxy diameters. A loose group (e.g., the Local Group; M81) contains on the order of 10 galaxies separated by 10-100 galaxy diameters. A rich cluster (e.g., Virgo; Coma) contains 100 or more galaxies within a volume comparable to that of a loose group. Scale of cluster, about 1 Mpc. 21 known X-ray sources are associated with clusters of galaxies.

cluster variable   See RR Lyrae star.

Coalsack   A prominent dark nebula in Crux, near the Southern Cross, readily visible to the naked eye, about 170 pc distant, located on the galactic plane.

coherence   The existence of a correlation (statistical or temporal) between the phases of two or more waves.

coherent scattering   A scattering process that leaves atoms in the same energy state after the scattered photon departs in a direction different from that of the incident photon. The energy of the scattered photon is the same (in the rest frame of the atom) as that of the incident photon.

cold-gas approximation (in MHD studies)   An approximation in which the sound speed is much less than the Alfvén speed or the gas pressure is much less than the magnetic pressure.

collapsed star   See black hole.

collinear (of three or more points)   Lying in a straight line.

color-color plot   Usually, plot of $B-V$ versus $U-B$.

color excess   Difference between the observed color index of a star and the intrinsic color index corresponding to its spectral type. It indicates the amount of reddening suffered by the light from the star when it passes through interstellar dust.

color index   Difference between the photographic and photovisual magnitudes of a star; or more generally, the difference in magnitudes between any two spectral regions. Color index is always defined as the short-wavelength magnitude minus the long-wavelength magnitude. In the Johnson-Morgan UBV system, the color index for an A0 star is defined as $B-V = U-B = 0$; it is negative for hotter stars and positive for cooler ones.

color-magnitude diagram (C-M diagram)   Plot of absolute or apparent visual magnitude against color index for a group of stars.

color temperature   A stellar temperature determined by comparison of the spectral distribution of the star's radiation with that of a blackbody.

column density (N)   The number of particles per square centimeter along a specified path with a length equal to the distance to the probing source.

coma   An aberration common in traditional reflecting telescopes, in which off-axis rays of light striking different parts of the objective do not focus in the same image plane. It produces elongated, cometlike images at the outer edge of the field. It is mainly because of coma that the Hale telescope is limited to on-axis work and has a usable field of only 10′ without special corrector lenses. This problem has largely been solved by the Schmidt telescope and the Ritchey-Chrétien design.

coma (of a comet)   The spherical region of diffuse gas, about 150,000 km in diameter, which surrounds the nucleus (q.v.) of a comet. Together, the coma and the nucleus form the comet's head.

Coma cluster (A1656)   A symmetric cluster of about 1000 galaxies (primarily E and S0) about 92 Mpc distant ($z = 0.023$). Luminous mass $4 \times 10^{14} M\odot = 8 \times 10^{47}$ g; virial theorem mass about 5×10^{48} g; mass needed to bind the cluster about 4×10^{49} g. $R \approx 9 \times 10^{24}$ cm. It is also an X-ray source (see Coma X-1).

Coma Cluster   An open cluster of about 100 stars in our Galaxy (about 80 pc distant). Similar to the Hyades in overall binary frequency.

Coma X-1 (3U 1257 + 28)   An extended X-ray source in the Coma cluster of galaxies.

combination variable   See symbiotic star.

comet   A diffuse body of gas and solid particles (such as CN, C_2, NH_3, and OH), which orbits the Sun. The orbit is usually highly elliptical or even parabolic (average perihelion distance less than 1 AU; average aphelion distance, roughly 10^4 AU). Comets are unstable bodies with masses on the order of 10^{18} g whose average lifetime is about 100 perihelion passages. Periodic comets comprise only about 4% of all known comets. Comets are obviously related in some manner to meteors, but no meteorites from a comet have ever been recovered. Observations of comets Bennett and Kohoutek have established that a comet is surrounded by a vast hydrogen halo.

comets, nomenclature   When a newly discovered comet is confirmed, the IAU assigns an interim designation consisting of the year of discovery followed by a lowercase letter in order of discovery for that year. Frequently the discoverer's name precedes the designation—e.g., comet Bennett 1969i. If a reliable orbit is later established, the comet is given a permanent designation consisting of the year of perihelion passage followed by a roman numeral in order of perihelion passage—e.g., comet Bennett 1970 II. If the comet is periodic, the letter P followed by the discoverer's (or computer's) name is used—e.g., comet 1910 II P/Halley.

comets, family of An aggregation of comets with similar aphelion distances (e.g., Jupiter's family).

comets, group of An aggregation of comets with identical orbits except for phase.

cometary nebula A reflection nebula with a fan shape that bears a superficial resemblence to a comet. Classical examples of the heads of cometary nebulae are R Mon, R CrA, and RY Tau. All have A0–G0 type spectra that resemble the spectrum of a T Tauri star, and their brightness varies from year to year.

commensurate orbits A term applied to two bodies orbiting around a common barycenter when the period of one is an integral multiple of that of the other.

compact galaxy A galaxy similar to an N galaxy but with no disk or nebulous background. It is an object of high surface brightness which appears slightly nonstellar on photographs and which has a larger redshift than normal stars in our Galaxy. Nearest "compact" galaxy is M32.

compact H II region A dense ($n_e \geq 10^3$ cm^{-3}) H II region of small linear dimensions (≤ 1 pc).

compact radio source A radio source which has a small angular extent and is strongest at shorter wavelengths (cf. extended source).

companion of Sirius (Sirius B) A white dwarf of about 1 solar mass but of only 0.03 solar radii ($R = 5400$ km, $T_{eff} = 32,000$ K).

comparison band The wavelength interval measured in the continuum outside a spectral feature—e.g., the 21-cm line.

Compton effect Decrease in the frequency of high-energy radiation (such as X-rays) caused when a photon loses some of its energy to a free electron by collision.

Compton scattering Scattering of a photon due to the Compton effect (see also noncoherent scattering).

configuration (of a set of mass points) All the data that refer to the location of each mass point in ordinary space.

configuration mixing   The superposition of a number of wave functions belonging to different configurations.

conjunction   See elongation.

conservative scattering   Scattering that occurs in the absence of absorption.

conservative system   A system in which energy is conserved; i.e., one in which there is no dissipation of energy.

conserved quantity   A quantity that remains unchanged in the course of the evolution of a dynamical system. There are seven known quantities that are conserved: energy (including mass), momentum, angular momentum (including spin), charge, electron-family number, muon-family number, and baryon-family number.

constants   Avogadro's number 6.02×10^{23}; 1 amu $= 1.66 \times 10^{-24}$ g; $m_e = 9.1 \times 10^{-28}$ g; $m_p = 1.00728$ amu; $m_H = 1.67 \times 10^{-24}$ g; $c = 299{,}792.46$ km s^{-1}; 1 AU $= 1.49598 \times 10^{13}$ cm; 1 lt-yr $= 9.4605 \times 10^{17}$ cm $= 6.324 \times 10^4$ AU; 1 lt-min ≈ 0.13 AU; 1 pc $= 3.084 \times 10^{18}$ cm $= 206{,}265$ AU $= 3.26$ lt-yr. $G = 6.668 \times 10^{-8}$ dyn cm^2 g^{-2}.

constellations

Abb.	Name	Genitive
And	Andromeda	Andromedae
Ant	Antlia	Antliae
Aps	Apus	Apodis
Aqr	Aquarius	Aquarii
Aql	Aquila	Aquilae
Ara	Ara	Arae
Ari	Aries	Arietis
Aur	Auriga	Aurigae
Boo	Bootes	Bootis
Cae	Caelum	Caeli
Cam	Camelopardalis	Camelopardalis
Cnc	Cancer	Cancri
CVn	Canes Venatici	Canum Venaticorum
CMa	Canis Major	Canis Majoris
CMi	Canis Minor	Canis Minoris
Cap	Capricornus	Capricorni
Car	Carina	Carinae
Cas	Cassiopeia	Cassiopeiae
Cen	Centaurus	Centauri
Cep	Cepheus	Cephei
Cet	Cetus	Ceti
Cha	Chamaeleon	Chamaeleontis

Cir	Circinus	Circini
Col	Columba	Columbae
Com	Coma Berenices	Comae Berenices
CrA	Corona Austrina	Coronae Austrinae
CrB	Corona Borealis	Coronae Borealis
Crv	Corvus	Corvi
Crt	Crater	Crateris
Cru	Crux	Crucis
Cyg	Cygnus	Cygni
Del	Delphinus	Delphini
Dor	Dorado	Doradus
Dra	Draco	Draconis
Equ	Equuleus	Equulei
Eri	Eridanus	Eridani
For	Fornax	Fornacis
Gem	Gemini	Geminorum
Gru	Grus	Gruis
Her	Hercules	Herculis
Hor	Horologium	Horologii
Hya	Hydra	Hydrae
Hyi	Hydrus	Hydri
Ind	Indus	Indi
Lac	Lacerta	Lacertae
Leo	Leo	Leonis
LMi	Leo Minor	Leonis Minoris
Lep	Lepus	Leporis
Lib	Libra	Librae
Lup	Lupus	Lupi
Lyn	Lynx	Lyncis
Lyr	Lyra	Lyrae
Men	Mensa	Mensae
Mic	Microscopium	Microscopii
Mon	Monoceros	Monocerotis
Mus	Musca	Muscae
Nor	Norma	Normae
Oct	Octans	Octantis
Oph	Ophiuchus	Ophiuchi
Ori	Orion	Orionis
Pav	Pavo	Pavonis
Peg	Pegasus	Pegasi
Per	Perseus	Persei
Phe	Phoenix	Phoenicis

Pic	Pictor	Pictoris
Psc	Pisces	Piscium
PsA	Piscis Austrinus	Piscis Austrini
Pup	Puppis	Puppis
Pyx	Pyxis	Pyxidis
Ret	Reticulum	Reticuli
Sge	Sagitta	Sagittae
Sgr	Sagittarius	Sagittarii
Sco	Scorpius	Scorpii
Scl	Sculptor	Sculptoris
Sct	Scutum	Scuti
Ser	Serpens	Serpentis
Sex	Sextans	Sextantis
Tau	Taurus	Tauri
Tel	Telescopium	Telescopii
Tri	Triangulum	Trianguli
TrA	Triangulum Australe	Trianguli Australis
Tuc	Tucana	Tucanae
UMa	Ursa Major	Ursae Majoris
UMi	Ursa Minor	Ursae Minoris
Vel	Vela	Velorum
Vir	Virgo	Virginis
Vol	Volans	Volantis
Vul	Vulpecula	Vulpeculae

continuous spectrum   An unbroken emission spectrum spanning the range of optical wavelengths from the infrared to the ultraviolet (or an unbroken emission band in the radio region). A continuous spectrum occurs when free electrons describing energetic orbits encounter an atomic nucleus, emit radiation, and drop into less energetic orbits (either bound or free). Since the energies of free electrons are not quantized, but are always greater than the energies of bound electrons, their emissions form a continuous spectrum in the spectral region beyond the series limit.

continuum   A set of points which form a line (one-dimensional continuum), a plane (two-dimensional continuum), etc. See also continuous spectrum.

Coordinated Universal Time (UTC)   Universal Time coordinated with ephemeris time; i.e., the rate is defined relative to atomic clock rate, but the epoch is defined relative to Universal Time. UTC is defined in such a manner that it differs from International Atomic Time (IAT) by an exact whole number of seconds. The difference UTC minus IAT was set equal to -10 s starting 1972

January 1; this difference can be modified by 1 s, preferably on January 1 and in case of need on July 1, to keep UTC in agreement with the time defined by the rotation of the Earth with an approximation better than 0.7 s (see atomic time).

Copernicus   An Orbiting Astronomical Observatory (OAO-3), launched 1972 August 21 ($a = 7123$ km, $e = 0.00083$, $i = 35°.0$) equipped with an ultraviolet telescope, a steerable X-ray telescope, and gamma-ray detectors.

coplanar   Lying in one plane.

core-halo galaxies   A class of radio sources characterized by an emission "halo" surrounding a more intense "core." About 20% of the known extended radio sources are of the core-halo type.

Coriolis effect   The acceleration which a body in motion experiences when observed in a rotating frame. This force acts at right angles to the direction of the angular velocity. Thus a projectile fired due north from any point on the northern hemisphere will land slightly east of its target because the eastward velocity of Earth's surface decreases from the equator to the poles. The Coriolis effect is responsible for large-scale wind patterns in Earth's atmosphere (and for ocean currents).

corona   Outermost atmosphere of the Sun immediately above the chromosphere, consisting of hot ($1–2 \times 10^6$ K), low-density (about 10^{-16} g cm^{-3}) gas that extends for millions of miles from the Suns's surface. Ordinarily it can be seen only during a total solar eclipse. Its shape varies from almost spherical at sunspot maximum to unsymmetrical at minimum. Its high temperature is probably caused by MHD shock waves generated below the photosphere. The corona, together with solar flares, is the source of solar X-rays. It is the corona, not the photosphere, that is studied by radio astronomers, except at very short wavelengths.

corona (or halo) of Galaxy   See halo.

R Coronae Borealis variables   A class of very luminous helium-rich, carbon-rich, hydrogen-poor eruptive variable supergiants. The prototype is R CrB, an F8–G0 Ib star with a large infrared excess and a rather high ^{7}Li abundance. It has fluctuated from 6th magnitude to 14th magnitude.

coronal equilibrium   The equilibrium in which collisional ionizations balance radiative recombinations.

coronal green line   An emission line of Fe XIV at 5303 Å—the strongest line in the solar corona.

coronal hole   An area where the extreme-ultraviolet and X-ray coronal emission is abnormally low or absent; a coronal region apparently associated with diverging magnetic fields. A great part, if not all, of the solar wind starts from coronal holes.

corpuscular radiation   Charged particles (mainly protons, alpha-particles, and electrons) emitted by a star (see also cosmic rays; solar wind).

correlator   In radio astronomy, an instrument which measures the similarity between the current fluctuations due to shot noise (q. v.) and those due to wave noise (q.v.).

cosecant law   The law of attenuation as one sees through a plane-parallel atmosphere. Thus the variation of the logarithm of the number of galaxies with galactic latitude b, by virtue of interstellar absorption, is log $\langle N \rangle$ = A − $0.6\Delta m$ csc b.

cosine law   See Lambert's law.

cosmic background radiation   Isotropic (to better than 1 part in 1000) radiation first detected in 1964 by Penzias and Wilson at a wavelength of 7.35 cm (T about 2.7 K). It has since been observed at radio wavelengths from 1 mm to 21 cm, and also at X- and gamma-ray frequencies. The cosmic background radiation is interpreted as relict radiation from the primeval fireball; it represents a z of approximately 3000 (see crossover time).

cosmic γ-ray bursts   Short (about 0.1–4 s), intense, low-energy (about 0.1–1.2 MeV) bursts, first recorded by the *Vela* satellite system on 1967 July 2, but not declassified until 1973. About five events per year have been detected. Their isotropic distribution suggests an extragalactic origin, but a galactic disk origin cannot be ruled out: there is a large increase in γ-ray flux in the direction of the galactic center.

cosmic light   A small (no more than 1%) contribution by extragalactic sources to the background glow of the night sky.

cosmic rays   High-energy charged particles (about 85% protons, 14% α-particles, 1% electrons, $\ll$1% heavy nuclei) which stream at relativistic velocities (mean energy $\sim$2 GeV) down to Earth from space. The Sun ejects low-energy ($10^7 - 10^{10}$ eV) cosmic rays during solar flares (those of lower energy than this are unobservable from Earth because of solar system magnetic fields). Those of intermediate energy ($10^{10} - 10^{16}$ eV) have an isotropic distribution, and are apparently produced in the Galaxy. Possible sources of acceleration are shock waves accompanying supernovae (although cosmic rays have a higher hydrogen content than would be expected from a star that has processed material to iron), and the rotating magnetic fields of pulsars. The light elements Li, Be, and B have a higher abundance ratio in cosmic rays than in the solar system.

cosmogony   The study of the origin of celestial systems, especially the solar system.

cosmological constant (Λ) (also called cosmical constant)   A term introduced by Einstein into his field equations to permit a static, homogeneous, isotropic model of the Universe. Λ may be taken to be positive, negative, or zero. If positive, it represents a repulsive force in the Universe and is directly proportional to the distance—the greater the distance between two galaxies, the greater the value of Λ.

cosmological distances   Distances implied by assuming the validity of the Hubble relation between redshift and distance.

cosmological hypothesis   The assumption that quasars are at distances inferred from their redshifts.

cosmological model   The result of the theoretical calculation of an expansion curve obtained, for the case of relativistic cosmological models, from solutions to Einstein's field equations. A cosmological model is intended to represent the positions and motions of the material in the universe.

cosmological principle   The assumption that all observers, everywhere in space, would at a given cosmic time view the same large-scale picture of the Universe (cf. perfect cosmological principle).

cosmological redshift   The redshift (q.v.) due to the expansion of the Universe. The redshift as observed is the ratio of the radius of the Universe at the present epoch to the radius of the Universe at the time the radiation left the distant object.

cosmology   The study of the origin, structure, and evolution of the Universe on the largest possible scale. In present usage, it frequently includes cosmogony (q.v.).

Coster-Kronig transition   An Auger transition (q.v.) in which the vacancy is filled by an electron from a higher subshell of the same shell.

coudé focus (the word comes from a French word meaning "bent like an elbow," not from a man's name)   A focus used primarily for spectroscopy. In this arrangement, light from the primary mirror is reflected along the polar axis to focus at a fixed place separate from the moving parts of the telescope, where large pieces of equipment can be fitted without interfering with the telescope's balance.

coulomb   The SI unit of charge. 1 coulomb = 2.998×10^9 esu.

Coulomb-Born approximation   An approximation similar to the Born approximation (q.v.) except that Coulomb waves replace plane waves for the incident and scattered photons.

Coulomb collision   The collision between two charged particles.

Coulomb's law   The force between two charged particles varies directly as the size of the charges and inversely as the square of the
distance between them.

counting rate   See proportional counter.

coupling   An interaction between the components of a system.

covariant   An adjective applied to a set of relationships between
mathematical or physical quantities if they remain unchanged
after transformation to a different coordinate system.

Crab Nebula (M1, NGC 1952)   A chaotic, expanding mass of gas in
Taurus, about 2 kpc distant, the remnant of a Type I supernova
whose light reached Earth in 1054. It is an intense radio source
(Tau A), and its visible light is strongly polarized. It is also a
source of X-rays (2U 0531 + 22) and gamma-rays. Its total mass
is about 1 $M\odot$, but the total energy radiated by the Crab is
10^{37}–10^{38} ergs s^{-1}. It is periodically occulted by the Moon, and
every June its radio spectrum is occulted by the solar corona.

Crab pulsar (NP 0532)   A pulsar associated with the Crab Nebula. It
has the shortest period (0.0331 seconds) of any known pulsar.

crepe ring   The second innermost ring (about 18,000–20,000 km wide)
of Saturn (see Saturn's rings). The crepe ring has fewer particles
and is less dense than the outer rings; therefore, it is harder to
observe. Discovered by Bond in 1850.

critical equatorial velocity   In rotating early-type stars, that velocity at
which the ratio of centrifugal force to gravity at the equator is
unity.

crossing time   The time it takes a particle to travel from one point in
its orbit to another point 180° away.

crossover effect   A term applied to the observation in magnetic stars
that line profiles are definitely sharper in circularly polarized
light of one sense than in that of the other. It often occurs when
the magnetic field changes sign.

crossover time   The epoch during the radiation era (q.v.) at $t \approx 10^{12}$
seconds, when the universe switched from being radiation-domi-
nated to being matter-dominated.

cross section   A measure of the probability that a given reaction will
occur. It is usually expressed in terms of an effective area that
the incident beam would have to hit to account for the reaction
rate.

cross-spectrum   The transform of the covariance spectrum.

culmination   The instant at which a celestial object crosses the meridian.

curie   Unit of radioactivity. 1 curie = 3.7 × 10¹⁰ disintegrations per second.

current density   Amount of charge passing through a unit area per unit time.

current sheath model   A sunspot model in which the cylinder of the magnetic field is assumed to be surrounded by a current sheath which contains all the gradients of the field.

curvature of spacetime   A notion associated with the description of spacetime in terms of Riemannian geometry. One distinguishes, in cosmological models, three types of space curvature: positive; zero (Euclidean); and negative (hyperbolic, geometry of Lobachevsky).

curve of growth   The relation between the equivalent width of an absorption line and the number of atoms that produce it.

cusp   A horn of the Moon, Mercury, or Venus when in the crescent phase.

cyanogen (CN) bands   Molecular bands found in the spectra of stars of type G0 and later. Cyanogen absorption is an important luminosity criterion, and is more pronounced in giants than in dwarfs of the same spectral type.

α Cygni   See Deneb.

P Cygni star   A type of star named after the fifth-magnitude B1e star P Cygni, about 1200 pc distant, whose spectrum shows strong emission lines, like those of the Be and Wolf-Rayet stars, with blueshifted absorption components which are presumed to come from an expanding shell of low-density matter. A P Cygni profile is taken as an indication of mass loss.

SS Cygni stars   A subclass of dwarf novae. SS Cyg is a double-lined, noneclipsing spectroscopic binary (sdBe, dG5) with an orbital period of 6^h38^m. Mean time between eruptions, 54 days. It may be a sporadic source of soft X-rays.

61 Cygni   A binary system 3.4 pc distant (parallax 0".293), consisting of a K5 and a K7 component with a period of about 720 years. One of the components is itself a binary with a period of about 5 years. It has an invisible component about 8 times the mass of Jupiter.

V444 Cygni   A close binary system (WN6 + B1) with a period of 4.21 days.

V1016 Cygni (= MHα 328-116)   A peculiar emission object (in optical, radio, and infrared), possibly a symbiotic star. (Perhaps 2 kpc distant?) It brightened visually by 4 mag in 1964–65. Before brightening, it was classified as a late M star. Its infrared variations are similar to those of Mira variables. It probably ejected a shell in 1964–65 and may be in the process of becoming a planetary nebula.

V1057 Cygni (= LkHα 190)   A former T Tauri star (spectral type K) which brightened in late 1969 by more than 5 mag. After the flare-up, the star had an A1-type spectrum (by 1972 its spectral type was back to F). It has many similarities to FU Orionis and is undoubtedly a pre-main-sequence star (M>2 $M\odot$). It is also an infrared emitter and an OH and CO source.

Cygnus A source (= 3C 405)   A double radio source, the third strongest radio source in the sky (after the Sun and Cas A), at one time believed to be caused by the collision of two galaxies. It has now been identified with a distant peculiar cD galaxy ($z \approx 0.056$). It is also an X-ray source (2U 1957+40).

Cygnus Loop   A gaseous nebula (a supernova remnant) (NGC 6992, Cyg X-5), consisting of a large loop of gas ejected from a star about 20,000 years ago. It is probably about 770 pc distant and 100 pc above the galactic plane. (X-ray observations give a distance of 2–3 kpc.) It is a thermal bremsstrahlung source of soft X-rays with a spectral temperature of 2×10^6 K.

NML Cygnus (IRC+40448)   An infrared star (heliocentric radial velocity -43 km s^{-1}) discovered by Neugebauer, Martz, and Leighton. Its spectral type is M6 III, and its surface temperature is about 700 K (about the surface temperature of Venus). It is a strong OH emitter, and CO has been identified in its spectrum. (Perhaps 200 pc distant?)

Cygnus X source   A complex of radio sources.

Cygnus X-1 (3U 1956+35)   An X-ray source about 2.5 kpc distant (period 5.5998 days, $e \approx 0.06$, $i \approx 27°$). The visible component is the ninth-magnitude supergiant HDE 226868 (O9.7 Iab). It has rapid night-to-night variations in spectral features. Mass of primary about 20 $M\odot$; mass of collapsed star, at least 6 $M\odot$.

Cygnus X-2 (3U 2142+38)   An X-ray source optically identified with an irregular variable star.

Cygnus X-3 (2U 2030+40)   An X-ray binary with a 4.8 hour period discovered in 1966. It is also an infrared source, a cosmic ray source, and a strongly variable radio source (interstellar extinction is too high for visible light observations). It is best fitted by a model of an expanding cloud of relativistic electrons emitting synchrotron radiation around a neutron star. About 10 kpc distant.

Cygnus X-5   See Cygnus Loop.

cynthion   Of or pertaining to the Moon. Diana, the Roman goddess of the Moon, was sometimes called Cynthia for her birthplace on Mount Cynthus in Delos.

Cytheran   Of or pertaining to the planet Venus. Cythera is the Ionian island on which the goddess Venus supposedly first set foot when she emerged from the foam.

D

***d*-electron**   An orbital electron whose *l* quantum number is 2.

D galaxy   A supergiant radio galaxy (the most common type of radio galaxy) which has an elliptical nucleus surrounded by an extended envelope. Or, an optical galaxy with a very bright nucleus. In the Morgan classification, a galaxy with rotational symmetry but without pronounced spiral or elliptical structure (a dustless galaxy). In the Yerkes 1974 system, a galaxy with an elliptical-like nucleus surrounded by an extensive envelope (see also R galaxy).

D layer   The lowest part of Earth's ionosphere (starts at about 100 km). This is the layer that reflects "broadcast" radio waves.

D lines   Two close spectral lines of neutral sodium (see Fraunhofer lines) at 5896 Å (D1) and 5890 Å (D2).

D ring   The innermost ring of Saturn, discovered in 1969.

damping   In any oscillating system, a decrease in the amplitude of an oscillation due to the dissipation of energy.

dark current   The current that flows in a photoelectric cell when not illuminated.

dark nebula   A relatively dense (up to 10^4 particles per cm^3) cloud of interstellar matter whose dust particles obscure the light from stars beyond it and give the cloud the appearance of a region devoid of stars.

Darwin ellipsoids   Ellipsoidal figures of equilibrium of homogeneous massive bodies describing circular orbits with a uniform angular velocity about each other on certain approximations relative to their mutual tidal influences.

db galaxy   One of a small number of dumbbell-shaped radio galaxies. They might be called D systems with double nuclei, in which two elliptical nuclei share a common extended envelope.

de Broglie wavelength (λ)   The wavelength associated with a particle with a momentum p: $\lambda = h/p$.

debye   Unit of electric-dipole moment, equal to that existing between a unit of positive charge and a unit of negative charge separated by a distance of 1 cm. 1 debye $= 10^{-18}$ statcoulomb cm.

Debye-Hückel model   The standard plasma model of an ionized classical gas.

Debye length   A theoretical length which describes the maximum distance that a given electron can be from a given positive ion and still be influenced by the electric field of that ion in a plasma. Although according to Coulomb's law oppositely charged particles continue to attract each other at infinite distances, Debye showed that there is a cutoff of this force where there are other charged particles between. This critical separation decreases for increased density. $l_D = (kT_e/4\pi n_e e^2)^{1/2}$.

deceleration parameter (q_0)   A dimensionless quantity describing the rate at which the expansion of the Universe is slowing down because of self-gravitation: it gives a measure of the matter density. In Friedmann's equation (which describes many cosmological models) $q_0 = -1$ indicates a steady-state universe, $q_0 < +\frac{1}{2}$ indicates an open universe, $q_0 = +\frac{1}{2}$ indicates a flat Euclidean universe, and $q_0 > \frac{1}{2}$ indicates a universe that is decelerating and will eventually contract. Sandage and Tammann (1975) obtain $q_0 = 0.10$ for $H_0 = 55$ km s^{-1} Mpc^{-1}.

declination (δ)   Angular distance north ($+$) or south ($-$) of the celestial equator to some object, measured in degrees, minutes, and seconds of arc along an hour circle passing through the object. Declination is analogous to latitude on the Earth's surface.

decoupling epoch (in the Universe)   The epoch at t $\approx 10^{13}$ s after the big bang ($T \approx 3000$ K) when matter and radiation decoupled.

degeneracy pressure   Pressure in a degenerate electron or neutron gas.

degenerate gas   A state of matter found in white dwarfs and other ultrahigh-density objects, in which the electrons follow Fermi-Dirac statistics. According to the classical laws of physics, the pressure of a gas is proportional to the temperature and the density. However, in 1926 Fermi and Dirac showed that if the density were high enough, departures from classical laws would occur, in that if at a given temperature the density is increased, the pressure increases more and more rapidly until it becomes independent of the temperature and is a function of the density only. When this point is reached, the gas is said to be degenerate.

degrees of freedom   Total number of values (or least number of independent variables) that must be specified to define a dynamical system completely.

Deimos The outer satellite of Mars, 12 × 13 km, P = 1.26 days; e = 0.003; inclination of orbit to planetary equator 1°6. Visual geometric albedo 0.06. *Mariner 9* has shown that both Phobos and Deimos are locked in synchronous rotation with Mars. Discovered by A. Hall in 1897.

delay time Time lapse between the time a signal (e.g., a radar beam) is propagated out to a distant object and the time it is received after the object bounces it back.

delta (δ) function Also called Dirac function. A function $\delta(x)$ defined by Dirac as having the property of being zero for all values of x except at zero, and also having the property that its definite integral from $-\infty$ to $+\infty$ is unity.

δ-ray A recoil electron ejected from an atom by an energetic charged particle. Delta-rays appear as branches in the main track of a cloud chamber.

Demeter Unofficial name for Jupiter X. P = 259.2 days, e = 0.12, i = 29°. Discovered by Nicholson in 1938.

Demon star See Algol.

Deneb (α Cygni) An A2 Ia supergiant about 430 pc distant at the head of the Northern Cross.

density wave A sound wave, or any other kind of material wave, which produces a series of alternate condensations and rarefactions of the material through which it passes.

density-wave theory A theory formulated about 1925 by Lindblad to explain the problem of spiral structure. Subsequent work by Lin and others attempts to explain the large-scale structure of spirals in terms of a small-amplitude wave propagating with fixed angular velocity. As the compression wave goes through, it triggers star formation on the leading edge of the spiral arms.

descending node The point in the orbit of a solar-system body where the body crosses the ecliptic from north to south.

de Sitter universe A geometrical model (1917) of an empty universe, based on Einstein's field equations.

detached binaries Binaries which are not in contact and in which no significant mass exchange is occurring.

deuterium (D) An isotope of hydrogen, ^{2}H (mass 2.0141 amu), discovered in interstellar space in 1965. The D/H number ratio in interstellar space is 1.4×10^{-5} (D/H on Sun, less than 10^{-6}, in Earth's oceans, 1.6×10^{-4}). Because deuterium is quickly destroyed in nuclear reactions, one view is that most of the D in the universe is primordial.

deuteron (d)   The nucleus of a deuterium atom. m_d = 2.01355 amu.

dex (abbreviation for decimal exponent)   A notation which converts the number after it into its common antilogarithm; e.g., dex (1.27) = $10^{1.27}$.

dielectronic recombination   Inverse autoionization.

differentiation (in a planet)   A process whereby the primordial substances are separated, the heavier elements sinking to the center and the lighter ones rising to the surface.

diffraction   The bending of light in passing a sharp edge or tiny aperture.

diffraction grating   A system of parallel slits, where the slit width is of the same order as the wavelength of the incident radiation, which is capable of dispersing light into its spectrum.

diffraction limited   Capable of producing images with angular separations as small as the theoretical limit.

diffraction pattern   A series of concentric rings produced by interference (q.v.).

diffuse galactic light   Scattered, integrated starlight; a small contribution to the background glow by starlight reflected and scattered by interstellar dust near the galactic plane. It is an extremely difficult quantity to measure and requires exceptionally dark sky conditions.

diffuse nebula   An irregularly shaped cloud of interstellar gas or dust whose spectrum may contain emission lines (emission nebula) or absorption lines characteristic of the spectrum of nearby illuminating stars (reflection nebula).

dilute aperture   In radio astronomy, an unfilled aperture.

dilution factor   Ratio of the energy density in a radiation field to the equilibrium value for radiation of some color temperature. $W = J_\nu / B_\nu (T_c)$. The dilution factor is of importance when the source of radiation subtends only a small angle, as seen from the absorbing gas.

Dione   Fifth satellite of Saturn. Radius about 440 km, P = 4.5 days. Discovered by Cassini in 1684.

dipole   A system composed of two equal charges of opposite sign, separated by a finite distance; e.g., the nucleus and orbital electron of a hydrogen atom. An ordinary bar magnet is a dipole.

dipole antenna   A type of array consisting of a system of dipoles often used with radio telescopes. It differs from the dish antenna in that it consists of many separate antennas that collect energy by feeding all their weak individual signals into one common receiving set.

Dirac function   See delta function.

direct motion   Motion of a solar-system body from west to east across the sky.

direction cosine   A means of specifying the direction of a celestial object. The direction cosine involves the inclination of a vector to each of three mutually perpendicular axes defined through a point in space.

dirty ice (interstellar)   Interstellar ice grains with graphite particles or other impurities adsorbed on their surfaces.

dish   A large parabolic "mirror" of sheet metal or wire mesh which collects radio energy and reflects it to the antenna at the focus.

disjoint   An adjective applied in mathematics to two or more sets which have no members in common.

disk (of a spiral galaxy)   The central plane, as distinguished from the halo or the nucleus.

disk star   A star located in the disk of the Galaxy.

dispersion (optical)   Resolution of white light into its component wavelengths, either by refraction or by diffraction.

dispersion (radio)   The selective retardation of radio waves when they propagate through an ionized gas. The speed of propagation depends on the frequency—the lower the frequency, the greater the time lag. Thus if a pulsar emits a sharp pulse of radio waves containing a spread of wavelengths, the waves will interact with the charged particles in the interstellar medium, and the higher frequencies will arrive at the Earth slightly before the lower frequencies.

dispersion measure   A term used in radio astronomy to describe the amount of dispersion in a radio signal. It is proportional to the product of the number of interstellar electrons per cm^3 and the distance to the source (in parsecs).

displacement   The distance (linear or angular) of an oscillating particle from its equilibrium position at any instant.

dissociative recombination   Capture of an electron by a positive molecular ion, wherein part of the recombination energy dissociates the molecule into two neutral atoms.

distance modulus   Difference between the apparent and absolute magnitudes: $m - M = 5 \log (r/10)$, where r is in parsecs. It is used to calculate the distance to a star.

distribution function   A function that gives the relative frequency with which the value of a statistical variable may be expected to lie within any specified interval (cf. statistical error). For example, the Maxwellian distribution of velocities gives the number of particles, in different velocity intervals, in a unit volume.

diurnal motion The apparent westward motion of celestial bodies, as seen from Earth, due to the Earth's axial rotation.

Doppler broadening The broadening of spectral lines caused by the thermal, turbulent, or mass motions of atoms along the line of sight. Small displacements of radiation absorbed or emitted by these atoms toward longer and shorter wavelengths result in broadening of the lines.

Doppler shift Displacement of spectral lines in the radiation received from a source due to its relative motion in the line of sight. Sources approaching ($-$) the observer are shifted toward the blue; those receding ($+$), toward the red. The Doppler shift makes it possible to determine the radial velocity and the rotation of stars.

S Doradus A supergiant eclipsing binary (an Eta Carinae-type object) in the Large Magellanic Cloud. It has a period of about 40 years.

30 Doradus Nebula (NGC 2070) A giant H II region, at least 300 pc across—one of the largest known—in the Large Magellanic Cloud. It is larger and more luminous than any known in the Galaxy. It is the brightest object ($M_V = -19$) in the Cloud at both optical and radio wavelengths, and contains the densest concentration of W-R stars. (The brightest object near the center is a O+ WN star of $M_V = -10.2$.) It is characterized by very rapid, disordered, and complex motions.

double cluster in Perseus See h and χ Persei.

double radio source A radio galaxy (q.v.), the bulk of whose radio emission comes from two sources on opposite sides of the visual galaxy. The radiation is presumably the result of an explosion in the nucleus of the parent galaxy, which caused the ejection at high speed of energetic particles in two opposite directions. About one-third of all known radio galaxies are double sources.

doublet In a spectrum, a pair of associated lines arising from transitions having a common lower energy level.

Draco system A dwarf elliptical galaxy, about 80 kpc distant, the intrinsically faintest ($M_v = -8.5$) member of the Local Group (next to And I–III). Diameter about 1 kpc.

α Dra See Thuban.

draconic month See nodical month.

drift curves In radio astronomy, the output response as a function of position for a given filter as the source passes through the beam (cf. velocity profile).

Dumbbell Nebula (M27, NGC 6853)   A planetary nebula of large apparent diameter and low surface brightness in Vulpecula about 220 pc distant.

duty cycle (of a beam)   The fraction of time a pulsed beam is on.

dwarf (d)   A main-sequence star (luminosity class V).

dwarf Cepheids (also called δ Scuti stars)   Type I Cepheids (spectral class A–F) ($\langle M_V \rangle$ +4 to +2) with periods of 1 to 3 hours.

dwarf galaxy   A galaxy with low luminosity.

dwarf nova (see also SS Cyg star, U Gem star)   A short-period binary system consisting of a hot white dwarf (or a hot blue sdBe subdwarf) and a much cooler and slightly more massive late-type main-sequence companion which fills its Roche lobe and is ejecting mass onto the white dwarf through its inner Lagrangian point. (The light from dwarf novae comes from four sources: a white dwarf, a cool main-sequence star, a hot spot, and a disk.) The outbursts are usually assumed to be caused by the explosive nuclear burning of hydrogen-rich material accreted onto the surface of a degenerate star.

dyad   An operator indicated by writing the symbols of two vectors without a dot or cross between them.

dynamical friction   That effect of stellar encounters (q.v.) which damps the initial motion of a star in the direction of its motion, in contrast to the diffusion process which randomizes it.

dynamical parallax   The "parallax" (i.e., distance) for a binary star whose orbit is well known, derived by using the mass-luminosity relation and Newton's generalization of Kepler's third law.

dyne   The cgs unit of force equal to the force necessary to give an acceleration of 1 cm sec^{-2} to a mass of 1 gram. 1 dyne of force is roughly equivalent to 1 mg of weight.

dyne centimeter   See erg.

E

E component   See L component.

e-folding time   The time within which the amplitude of an oscillation increases or decreases by a factor e ($e = 2.718. . . .$).

E galaxy   In both Hubble's and Morgan's classifications, an elliptical galaxy.

E layer (also called Kennelly-Heaviside layer)   The part of Earth's ionosphere (about 150 km) where the temperature gradient reverses and starts to rise. It reflects "short-wave" radio waves.

E line   A Fraunhofer line (q.v.) at 5270 Å. It is a blend of Fe I and Ca I.

e-process   A hypothetical group of nuclear reactions by which the iron group is assumed to be synthesized. At temperatures $> 5 \times 10^9$ K and densities $> 3 \times 10^6$ g cm^{-3} there are great numbers of collisions between high-energy photons and nuclei. These collisions break up the nuclei, the fragments of which promptly combine with other particles. Thus, there is in effect an equilibrium between formation and breakup. Since the iron group has the largest binding energies, the particles over the long run will tend to be trapped in these nuclei. The e-process (the e stands for equilibrium) is presumed to occur in a supernova explosion.

early-type emission stars   See Be stars.

early-type spiral   In Hubble's classification, a spiral with a large nuclear bulge and closely coiled arms.

early-type stars   Hot stars of spectral types O, B, A, and early F.

Earth   Third planet from the Sun. Mass 5.977×10^{27} g; polar radius 6356.9 km; equatorial radius 6378.17 km; mean density 5.517 g cm^{-3}; effective temperature 287 K. Rotation period $23^h56^m4\overset{s}{.}1$. Mean distance from Sun 149,598,500 km (8.3 lt-min); perihelion distance (early January) 147,100,000 km; aphelion distance 152,100,000 km; v_{orb} 29.78 km s^{-1}; orbital period $365\overset{d}{.}2564$; $e = 0.0167$, $i = 0$; obliquity (1973) $23°26'34''$. Albedo 0.39 (water and land about 0.2; snow and clouds about 0.8). Surface gravity 980 cm s^{-2}; V_{esc} 11.19 km s^{-1}. Precession $50\overset{''}{.}256$ per year; relativistic advance of perihelion $4\overset{''}{.}6$ per century. Atmosphere (by volume) 78% N_2; 20.9% O_2; 0.9% Ar, 0.03% CO_2. Particle density of Earth's atmosphere at sea level 10^{19} per cm^3 = 1.3×10^{-3} g cm^{-3}. Atmospheric pressure at sea level 1.013×10^6 dyn cm^{-2}. Magnetic field at surface, about 0.5 gauss; in core, about 100 gauss. Core temperature about 6400 K; core density about 10 g cm^{-3}. Age $4.6 \pm 0.1 \times 10^9$ years. First forms of life appeared about 3.2 to 3.5×10^9 years ago (*Homo sapiens* appeared as a species about 10^5 years ago).

earthlight   Light reflected from the Earth's atmosphere onto the dark part of the Moon.

east point   The point on the celestial horizon 90° clockwise from the north point. At the equinoxes the Sun rises in the east point.

eccentricity (of an elliptical orbit) The amount by which the orbit deviates from circularity: $e = c/a$, where c is the distance from the center to a focus and a is the semimajor axis.

eclipse The total or partial obscuration of the light from a celestial body caused by its passage into the shadow of another body (cf. occultation).

eclipse year The interval of time (346.62 days) between two successive passages of the Sun through the same node of the Moon's orbit.

eclipsing binary See photometric binary.

ecliptic Plane of the Earth's orbit. (Strictly speaking, the ecliptic is a mathematical fiction corresponding not to the actual plane of the Earth's orbit, but to one with all minor irregularities smoothed out.)

Eddington approximation An approximation used in the study of radiative transfer. It is the assumption that the ratio of the second moment of the radiation field to the mean intensity is everywhere equal to ⅓, the value of this ratio for an isotropic field.

Eddington limit In essence, radiation pressure must not exceed gravity. It is the limit beyond which the radiation force on matter in the emitting region is greater than the gravitational forces that hold the star together. $L_E = 4\pi cGM/K_s$, where K_s = Thomson and/or Compton scattering opacity. Eddington limit for a 1 $M\odot$ star, 10^{38} ergs s^{-1}.

Eddington's standard model A stellar model in which energy is transported by radiation throughout the whole star and the ratio of the radiation pressure to the gas pressure is assumed to be constant.

edge effects Absorption in the spectra of galaxies at the edges of some passbands by lines broadened by velocity dispersion.

effective radius (of a galaxy) The distance from the center within which half the luminosity is included (cf. Holmberg radius).

effective temperature (T_{eff}) The temperature that a blackbody would have which emitted the same amount of energy per unit area as the star does; it is a temperature characteristic of the surface region. T_{eff} of Sun, 5800 K.

eigenfunctions The wave functions (q.v.) corresponding to the eigenvalues. Eigenfunctions represent the stationary states ("standing waves") of a system.

eigenvalues   The specified values of E (quantized energy) for which the Schrödinger wave equation is soluble, subject to the appropriate boundary conditions.

eikonal approximation   An approximation in which the oscillation of a wave front is replaced by the direction of the ray which is normal to the oscillation.

Einstein coefficient   An emission (or absorption) coefficient. A_{ji} is the coefficient of spontaneous emission; B_{ji} is the coefficient of stimulated emission, where i is the lower level and j is the upper level.

Einstein–de Sitter model   A homogeneous world model (the simplest relativistic model) of finite density, zero curvature, and nonzero cosmological constant, subject to the field equations of general relativity in an expanding Euclidean space. The radius increases rapidly from zero, and, although it always increases, the rate of increase becomes less as time goes on.

Einstein effect   Displacement of spectral lines due to the gravitational redshift.

Einstein universe   A world model of a static universe with a positive cosmological constant, whose radius of curvature is constant and independent of time.

Ekman layer   Upper boundary layer within which the amplitude changes exponentially.

elastic collision   A collision between two particles which conserves the total kinetic energy and momentum of the system. For atomic collisions it is one involving energy less than the excitation potential of the atom.

electromagnetic unit (emu)   A system of electrical units based on the electromagnetic properties of an electric current.

electron   A stable, negatively charged elementary particle—the lightest massive particle known. The classical electron radius is 2.82×10^{-13} cm; $m_e = 9.1 \times 10^{-28}$ g $= 5.48597 \times 10^{-4}$ amu. The electron family (see lepton) includes the electron e^-, the positron e^+, the electron neutrino ν_e, and the electron antineutrino $\bar{\nu}_e$. Rest-mass energy of electron 8.186×10^{-7} ergs. Electron charge $= 1.60219 \times 10^{-19}$ coulombs.

electron conduction (in astrophysics)   A process occurring in highly ionized stellar interiors where the density is high, whereby the bulk of the energy is transported by "hot" electrons moving in one direction and cooler electrons in the other. In degenerate matter electron conduction, not radiation, is the main mechanism of energy transport.

electron temperature   The temperature that appears in the Maxwell distribution of velocities for electrons.

electron-phonon scattering   Electron scattering by ions oscillating about equilibrium positions which form a perfect lattice.

electon volt (eV)   A unit of energy used in atomic and nuclear physics; the kinetic energy acquired by one electron in passing through a potential difference of 1 volt in vacuum. 1 eV = 1.60 × 10^{-12} ergs. An electron with an energy of 1 eV has a velocity of about 580 km s^{-1}. The wavelength associated with 1 eV is 12,398 Å.

electrostatic unit (esu)   A unit of charge defined as the charge which exerts a force of 1 dyne on a charge of equal magnitude at a distance of 1 cm.

elements (of an orbit)   Seven quantities that must be established from observations in order to define the size, shape, and orientation of an orbit in space. The six elements needed to determine the orbit of a solar-system body are the semimajor axis a (in AU), the eccentricity e, the inclination i of the object's orbital plane to the ecliptic, the longitude Ω of the ascending node, the argument ω of the perihelion, and the epoch T (usually the time of perihelion passage). To determine the orbit of a binary star system in which the mass is not known, a seventh element, the period, must be established.

elliptical galaxy   A galaxy without spiral arms and with an ellipsoidal shape. Ellipticals have little interstellar matter and no blue giants—the only giants are red, and they give ellipticals a slightly redder color than spirals. The most massive galaxies known (about 10^{13} M ⊙), as well as some of the least massive known, are ellipticals. No giant elliptical is near enough for any individual stars to be resolved. Ellipticals apparently produce only Type I supernovae.

elongation (of a planet)   The angle planet-Earth-Sun. Eastern elongations appear east of the Sun in the evening; western elongations, west of the Sun in the morning. An elongation of 0° is called conjunction; one of 180° is called opposition; and one of 90° is called quadrature.

emersion   The reappearance of a celestial body after eclipse or occultation.

emission coefficient   Radiant flux emitted per unit volume per unit solid angle.

emission measure (EM)   The product of the square of the electron density times the linear size of the emitting region (in parsecs).

emission nebula   An H II region whose spectrum consists of emission lines.

emission spectrum   A spectrum consisting of emission lines produced in the laboratory by a glowing gas under low pressure.

Enceladus   Third satellite of Saturn, about 500 km in diameter. Orbital period 1.37 days. Discovered by Herschel in 1789.

Encke's comet   The comet with the shortest known period (3.30 years) ($a = 2.21$ AU, $e = 0.847$, $i = 12°.4$). It has been observed at every apparition since its discovery in 1819. Its period is gradually decreasing. Named after J. F. Encke, who computed its orbit. (It was discovered by Pons.)

Encke's division   A region of decreased brightness in the outermost ring of Saturn.

encounter   See gravitational encounter.

endoergic process   A process in which some of the energy of the incoming particle is transferred to the nucleus.

endothermic   An adjective applied to a reaction in which a net input of energy is required for the reaction to occur.

energy curve   A plot of the intensity of the continuous spectrum versus the wavelength.

energy distribution   The amount of energy radiated at each range of wavelengths.

energy level   Any of the several discrete states of energy in which an atom or ion can exist. For example, an orbital electron can exist only in those energy levels that correspond to an integral number of deBroglie wavelengths in a Bohr atom.

energy spectrum (in cosmic-ray studies)   Plot of number of particles versus energy.

ensemble average   An average over an ensemble of all possible systems.

enthalphy (H)   The heat content of a body. $H = U + pV$, where U is the internal energy, p is the pressure, and V is the volume.

entropy   A measure of the amount of unavailable heat in a system; or a measure of the amount of disorder in a system.

Eötvös experiment   An experiment performed in 1909 by the Hungarian physicist Eötvös to establish that the gravitational acceleration of a body does not depend on its composition—i.e., that inertial mass and gravitational mass are exactly equal.

Ep galaxy   In Morgan's classification, an elliptical galaxy with dust absorption.

ephemeris (pl., ephemerides)   A list of computed positions occupied by a celestial body over successive intervals of time.

ephemeris second The length of a tropical second (1/31,556,925.97474 of the tropical year) on 1900 January 0.5 ephemeris time.

ephemeris time Time based on the ephemeris second. Ephemeris time is determined primarily from observations of the Moon against the background of stars, whereas Universal Time is determined from observations of the stars and depends on the Earth's current rate of rotation.

epoch A point of time selected as a fixed reference.

equation of state A relation between the pressure, temperature, and density of a fluid.

equation of time The difference between apparent and mean solar time. At Greenwich, apparent solar noon varies between $11^h44^m05^s$ and $12^h14^m19^s$. Maximum contribution from Earth's orbital eccentricity, ~ 8 min; from Earth's obliquity, ~ 10 min. Apparent and mean solar time agree 4 times a year.

equilibrium A condition of balance between the forces operating on or within a physical system, so that no accelerated motions exist among the parts of a system. For stable equilibrium, a small disturbance will eventually damp out. If a small disturbance continues to grow, the system is said to be in unstable equilibrium.

equilibrium position The position of an oscillating body at which no net force acts on it.

equinox Either of the two points on the celestial sphere where the celestial equator intersects the ecliptic.

equipartition of energy Equal sharing of the total energy among all components of a system. The principle originally enunciated by Boltzmann states that the mean kinetic energy of the molecules of a gas is equally divided among the various degrees of freedom possessed by the molecules. The average molecular energy associated with any degree of freedom is one-half the product of the absolute temperature and Boltzmann's constant.

equivalence, principle of The principle that it is impossible to distinguish between gravitational and inertial forces; gravitational mass is precisely equal to inertial mass.

equivalent width A measure of the total amount of energy subtracted from the continuous spectrum by an absorption line on a graph of relative intensity versus wavelength. Since the shapes of line profiles vary—e.g., one may be broad and shallow whereas another is narrow and deep—measurement is facilitated by transforming each profile into a rectangle whose base corresponds to zero intensity and whose area is the same as that of the true absorption line.

erg (sometimes called dyne cm)   The cgs unit of energy; the work done by a force of 1 dyne acting over a distance of 1 cm. 1 erg $= 10^{-7}$ joules $= 1$ g cm^2 s^{-2}.

ergodic motion   Motion by one or more particles which fills phase space with uniform density after a sufficiently long time.

ergoregion   That part of space in which no physical object can remain at rest with respect to an observer at infinity; the dragging of inertial frames is so extreme that all timelike world lines rotate with the star. Technically, it is the region in which the asymptotically timelike Killing vector becomes spacelike.

ergosphere   The region surrounding the event horizon (but inside the stationary limit) of a rotating Kerr black hole (see ergoregion).

α Eri   See Achernar.

AS Eri   An eclipsing binary whose secondary is close to its Roche limit.

ϵ Eri   A fourth-magnitude K2 V star 3.30 pc distant. In 1973 van de Kamp announced that it has a planet-like object in orbit around it at a distance of about 8 AU and with a period of about 25 years.

40 Eridani   A nearby triple system, 5 pc distant. Component A is K0 V; component B is a DA white dwarf; component C is M5e V.

Eros   A small asteroid, No. 433 (axes $35 \times 16 \times 17$ km) whose closest approach to Earth is less than 0.15 AU. Rotation period $5^h16^m12^s913$, orbital period 642 days, $a = 1.48$ AU, $e = 0.223$, $i = 10°.8$; perihelion distance 1.084 AU. Discovered by G. Witt in 1898.

eruptive galaxy   See violent galaxy.

eruptive variable   See cataclysmic variable.

escape velocity   The velocity that a body requires to achieve a parabolic orbit around its primary $[V_e = \sqrt{(2GM/R)}]$. Escape velocity at Earth's surface is 11.2 km s^{-1}; of Moon, 2.4 km s^{-1}; of Sun, 617.7 km s^{-1} (cf. orbital velocity).

Europa (J II)   One of the Galilean satellites of Jupiter, 3600 km in diameter. Period 3.55 days, $e = 0.00$, $i = 0°.01$, mean density 3.07 g cm^{-3}.

evection   The small irregularity in the Moon's orbital motion due to solar and planetary perturbations.

even-even nuclei   See 4N nuclei.

even-odd nuclei   Nuclei that contain even numbers of protons but odd numbers of neutrons.

event   A "point" in four-dimensional spacetime.

event horizon   The surface of a black hole: the boundary of a region in space from which no matter can escape and hence from which no signals can be received by an external observer (cf. stationary limit). Bodies inside event horizons disappear not temporarily but for all time. In the Schwarzschild geometry, the event horizon and the stationary limit coincide.

Evershed effect   The radial motion outward (from the central umbra) of the gases in the penumbral regions of sunspots.

exchange correlation   The correlation of particles and spins which is embodied in a Slater-determinant wave function.

excitation potential   Amount of energy required to bring an electron from its ground state to a given excited state (measured in electron volts).

exclusion principle (Pauli exclusion principle)   No two electrons in an atom can have the same set of values for the four quantum numbers n, l, m_l, m_s. The exclusion principle applies only to fermions, not to bosons.

exoergic process   A process in which energy is liberated.

"expanding arm"   A spiral arm of neutral hydrogen lying between 2.5 and 4 kpc beyond the galactic center and receding from it at about 135 km s^{-1}.

exploding galaxy   See violent galaxy.

explosive nucleosynthesis   The nucleosynthetic processes which are thought to occur in supernovae. These explosive processes are thought to produce the nuclei from neon up to and including the e-process nuclei (q.v.) and possibly the r-process nuclei (q.v.). Explosive carbon burning occurs for a temperature of about 2 $\times$ 10^9 K and a density of 10^4 $-$ 10^7 g cm^{-3} and produces nuclei from neon to silicon. Explosive oxygen burning occurs for a temperature of about 4 $\times$ 10^9 K and produces nuclei from silicon to calcium, and the e-process occurs at a temperature greater than 5 $\times$ 10^9 K and produces the iron peak nuclei.

explosive variables   See cataclysmic variables.

extended source   In radio astronomy, formerly a source whose angular extent could be measured, as distinguished from a point source. Now, one which has a large angular extent and is strongest at longer wavelengths (distinguished from a compact source). Most extended sources tend to be polarized.

extinction   Attenuation of starlight due to absorption and scattering by Earth's atmosphere, or by interstellar dust. The longer the path through the dust, and the denser the dust, the more the starlight is reddened. The normal relation is $A_V = 0.8$ mag per kpc. The total visual extinction toward the galactic center is on the order of 25 magnitudes.

F

F component (the F stands for Fraunhofer)   The outer part of the solar corona (see K component) which emits a continuous spectrum in which absorption lines can be seen. The F corona is caused by radiation from the photosphere scattered by interplanetary dust, and it decreases slowly with distance from the Sun.

F corona   See F component.

FDS law   See Fermi-Dirac-Sommerfeld law.

f-electron   An orbital electron whose l quantum number is 3.

F layers (also called **Appleton layers**)   Two layers in the Earth's ionosphere (F_1 and F_2 at about 200 and 300 km, respectively) immediately above the E layer.

f number (of a lens)   Ratio of the focal length to the diameter.

F region   Region of the ionosphere above the F layers.

f-spot   See sunspots.

F star   A star of spectral type F with a surface temperature of about 6000–7500 K, in which lines of hydrogen and Ca II are of about equal strength. Metal lines also become noticeable. Examples are Canopus, Procyon.

f-sum rule   The sum of the f-values for all the transitions from a given state (positive for absorption and negative for emission) is unity.

f-values   See oscillator strengths.

ft-values t = half-life of the β-unstable nucleus, and f stands for an integral which depends on the β-decay energy and the type of transition.

Fabry-Perot interferometer   A type of interferometer wherein the beam of light is passed through a series of pairs of partly reflecting surfaces set at various angles to it and spaced at certain prechosen numbers of the wavelength to be examined. It differs from the Michelson interferometer in that it has only one "arm."

faculae   Bright regions of the photosphere seen in white light, visible only near the limb of the Sun.

fall   A "fall" as opposed to a "find" is a meteorite whose arrival on Earth is witnessed. Stones constitute 92% of the observed falls.

far field   The field of a pulsar beyond the velocity-of-light radius.

Faraday effect   An effect occurring in H II regions in which a magnetic field causes a change in the polarized waves passing through (see Faraday rotation).

Faraday rotation   Rotation of the plane of polarization of linearly polarized radiation when the radiation passes through a plasma containing a magnetic field having a component in the direction of propagation.

Feautrier's method   A difference-equation method of solving transfer equations.

Fechner's law   The intensity of a sensation increases as the logarithm of the stimulus. (See Pogson's ratio.)

feed horn   A device used on a radio telescope. It is located at the focal point and acts as a receiver of radio waves which the antenna collects and focuses on it. It couples the energy into the lines to go into the amplifier.

femto   A prefix meaning 10^{-15}.

fermi   A unit of length equal to 10^{-13} cm.

Fermi-Dirac nuclei   Nuclei of odd A-number (i.e., nuclei that do not have integral spin) (cf. Bose-Einstein nuclei). Fermi-Dirac nuclei therefore obey the exclusion principle (q.v.).

Fermi-Dirac-Sommerfeld law   A law which gives the algebraic number of a quantized system of particles which have velocities within a small range.

Fermi gas   A gas of electrons (or, more generally, fermions) in which all the lowest quantum states are occupied. For such a gas, the pressure in the nonrelativistic limit is proportional to the 5/3 power of the density.

Fermi interaction   A weak interaction (q.v.) causing beta-decay.

fermion   An elementary particle whose spin is a half-integral multiple of $h/2\pi$. Fermions include the baryons, the leptons, and their antiparticles, and obey the Pauli exclusion principle (cf. boson).

fibrilles   Striations or streaks which are observed to form whirls in the solar chromosphere.

field curvature   An aberration in an optical instrument, common in Schmidt telescopes, in which the light rays come to a focus on a curved surface instead of on a plane.

field equations   Equations which relate to one of the fundamental fields of force. General relativity is called a field theory because it describes the gravitational field, and Einstein's equations of general relativity are called field equations. Maxwell's field equations describe the electromagnetic field.

field galaxy   An isolated galaxy which does not belong to any cluster of galaxies. The ratio of galaxies in clusters to field galaxies is about 23:1.

field stars   Stars distributed at random in space and not belonging to any particular star cluster.

figure of merit   The extent to which an optical system falls short of perfection.

filament   A prominence seen in projection on the solar disk.

filar micrometer   An instrument used with a telescope for measuring small angular separations (as between binary stars).

fine structure   Splitting of spectral lines by the spin-orbit energy—i.e., the potential energy of the inherent electron magnetic moment in the atom's own magnetic field.

fine-structure constant (α)   A "coupling constant," $e^2/\hbar c$, approximately $1/137$, that measures the strength of an interaction between a charged particle and the electromagnetic field. It gives a rough measure of the relative importance of relativistic and spin effects in the spectra of atoms.

fines, lunar   Small particles of rock or powdered rock on the Moon.

fireball   See meteor; see also primeval fireball.

First Point of Aries   See vernal equinox.

five-minute oscillations   Vertical oscillations of the solar atmosphere with a well-defined period of 5 minutes.

flare, solar   Sudden and short-lived (as short as 300 s) brightening of a region of the solar chromosphere in the vicinity of a sunspot, caused by the sudden release of large amounts of energy (up to 10^{32} ergs) in a relatively small volume above the solar "surface." During an intense solar flare (electron density 10^{11} compared with 10^8 in solar quiet times) the ionization in Earth's atmosphere may increase by several orders of magnitude. Solar flares are classified on a scale of importance ranging from 3^+ (largest area) to 1^- (smallest area). The largest solar flares eject a mass of about 10^{16} g at a speed of roughly 1500 km s^{-1}.

flare star (sometimes called **UV Ceti star**)   A member of a class of dwarf stars (usually dM3e–dM6e) that show sudden, intense outbursts of energy. The flares are usually rare and very short with mean amplitudes of about 0.5–0.6 mag. All known flare stars are intrinsically faint and have emission lines of H I and Ca II. It is generally believed that flares in flare stars have certain properties in common: rapid rise to peak light followed initially by a rapid decline and later by a slower phase that occasionally does not return to a preflare level within practical monitoring times (several hours). An increase in radio emission is often detected simultaneously with the optical outburst. About 30 flare stars are known, all within 20 pc. (In at least one theory, the flare star stage directly follows the T Tauri stage.)

flash spectrum   An emission spectrum of the solar chromosphere, obtained by placing an objective prism in front of the telescopic lens the instant before (or after) totality in a solar eclipse.

flickering   Aperiodic behavior in an oscillating system; rapid, large-amplitude variations in light.

flocculus   See plage.

fluorescence   The absorption of a photon of one wavelength and reemission of one or more photons at longer wavelengths, especially the transformation of ultraviolet radiation into visible light.

flux   Total radiation going out from the 2π solid angles of a hemisphere. If the radiation is uniform, the flux is 2π times the intensity (q.v.). Measured in ergs cm^{-2} s^{-1}.

flux density   Flux of radiation through a unit surface; the strength of an electromagnetic wave, defined as the amount of power incident per unit area. In radio astronomy, the brightness temperature integrated over the solid angle of the source yields the flux density.

flux tube   A tube of magnetic field lines.

flux unit   Unit of flux density. 1 f.u. = 10^{-26} watts per square meter per hertz (see jansky).

Fokker-Planck equation   A modified form of the Boltzmann equation allowing for collision terms in an approximate way. It is used in the problem of charged-particle transport in fluctuating electromagnetic fields.

Fomalhaut (α PsA)   An A3 V star about 7 pc distant.

forbidden lines   Spectral lines emitted from metastable states (q.v.), or those which have a very low probability (10^{-9}–10^{-10}) of occurrence. They appear at particle densities $\leq 10^8$ cm^{-3}. All forbidden lines have low excitation potentials. Forbidden lines are designated by enclosing them in brackets, e.g., [O II].

Forbush decrease   A decrease in cosmic-ray intensity with an *increase* in solar activity (and vice versa). This phenomenon was first noted by Forbush in 1954.

formaldehyde (H_2CO)   An organic molecule, the first polyatomic molecule to be discovered in interstellar space (in 1969). In 1973 it was discovered in two external galaxies.

formamide ($HCONH_2$)   A molecule discovered in interstellar space in 1971 at 6.5 cm.

formic acid (H_2CO_2)   A simple organic acid, the first to be detected in interstellar space (in 1970 at 18.3 cm). Formic acid is the "sting" of an insect.

Fornax A   A 10th-magnitude S0 galaxy (NGC 1316), which is a strong radio source.

Fornax system   A dwarf spheroidal galaxy, about 190 kpc distant, in the Local Group ($M_V \approx -12$, $M \approx 2 \times 10^7 \, M\odot$.

fossil Strömgren sphere   A relict H II region which remains after the evolution of its exciting star.

4-kpc arm   A component of the Sagittarius arm with noncircular gas motions (see 3-kpc arm).

4N nuclei (or even-even nuclei)   Nuclei possessing equal and even numbers of neutrons and protons. 4N nuclei are formed in supernova envelopes at temperatures of at least 2×10^9 K and are very stable.

four-vector (four-dimensional vector)   A quantity that has four components which, under the Lorentz transformation, transform like space and time. Instead of locating a point in three-dimensional space, a spacetime four-vector locates a point in four-dimensional spacetime.

Fourier analysis   The analysis of a periodic function into its simple harmonic components.

Fourier theorem   Any finite periodic motion may be analyzed into components, each of which is a simple harmonic motion of definite and determinable amplitude and phase.

frame of reference   A set of axes to which positions and motions in a system can be referred.

Franck-Condon principle   A theoretical interpretation of the relative intensity of vibrational transitions in an electronic band on the assumption that the intense transitions correspond to situations where an endpoint in the lower vibrational level is vertically below the corresponding endpoint in the upper vibrational level.

Fraunhofer lines   Absorption lines in the spectrum of the Sun, studied by Fraunhofer in 1814. The nine most prominent he labeled with capital letters (from the red end) A. B. C. D. E. F. G. H. and K. The A and B bands (at 7600 and 7100 Å) are now known to be groups of telluric lines due to O_2 absorption in Earth's atmosphere, and C and F are respectively known as Hα and Hβ.

free (of a particle)   Not bound to a nucleus.

frequency (ν)   Number of oscillations per second of an electromagnetic wave. The amplitude of a wave depends on the intensity; the wavelength, on the frequency.

frequency distribution   A statistical arrangement of numerical data according to size or magnitude (see also distribution function).

Friedmann universe   A homogeneous, isotropic model of the Universe involving nonstatic (i.e., expanding or contracting) solutions to Einstein's field equations (with zero cosmological constant) calculated by the Russian mathematician A. Friedmann in 1922.

fringes (interference)   Successive dark and light lines, caused by light beams that are out of phase alternating with those that are in phase. Interference fringes can be used to calculate the apparent diameter of a star.

Froissart bound   If in a hadron-hadron collision the absorption is complete, then the interaction radius cannot increase faster than the logarithm of the energy.

frozen-in   An adjective which applies to the abundance of elements produced by the *r*-process in a supernova when the temperature has dropped below the point at which they can serve as seed nuclei for further nucleosynthesis. It also applies to the magnetic field lines of a star in gravitational collapse.

full width at half-maximum (FWHM)   The full width of a spectral line at half-maximum intensity.

fundamental frequency   The lowest characteristic frequency of oscillation of a dynamical system.

fundamental stars   Stars for which coordinates have been determined to a very high degree of accuracy.

funneling   The concentration of stars from different parts of the main sequence in the red-giant region.

future light cone   See light cone.

G

G-band   A band of CH at 4303 Å. It is conspicuous in the spectra of G–K stars.

g-factor   Ratio of a particle's magnetic moment to its spin angular momentum.

gf-values   Weighted oscillator strengths. f = oscillator strength of the transition; g = statistical weight of the lower level.

G star   Stars of spectral type G are yellowish stars with surface temperatures of about 5000 to 6000 K, in which the H and K lines of Ca II have become dominant and in which a tremendous profusion of spectral lines of both neutral and ionized metals, particularly iron, begins to show. The Balmer lines of hydrogen are still recognizable. Examples are the Sun and Capella.

galactic cluster   See open cluster.

galactic coordinates   A system of coordinates based on the mean plane of the Galaxy, which is inclined about 63° to the celestial equator. Galactic latitude (b) is measured from the galactic equator north ($+$) or south ($-$); galactic longitude (l) is measured eastward along the galactic plane from the galactic center. In 1958, because of increased precision in determining the location of the galactic center, a new system of galactic coordinates was adopted, with the origin at the galactic center in Sagittarius at $\alpha(1950) = 17^h42^m4$, $\delta(1950) = -28°55'$. The new system is designated by a superior roman numeral II (i.e., b^{II}, l^{II}) and the old system by a superior roman numeral I: $l^{II} \approx l^I + 32°31$. Galactic coordinates are independent of precession.

galactic equator   The primary circle defined by the central plane of the Galaxy.

galactic light   See diffuse galactic light.

galactic poles   The poles of the galactic plane. The new system puts the galactic north pole in Coma at $\alpha(1950) = 12^h49^m$, $\delta(1950) = 27°4$.

galactic wind   A hypothetical outflow of tenuous material from a galaxy, analogous to the solar wind.

galaxy   A large (10^8–10^{13} $M\odot$), gravitationally bound aggregate of stars and interstellar matter. Galaxy formation is currently believed to have occurred around $z \approx 3$–4.

Galaxy   The galaxy to which the Sun belongs. Our Galaxy is about 10^{10} years old and contains about 10^{11} stars. Its mass is at least 10^{11} $M\odot$, about 5–10 percent of which is in the form of gas and dust. Diameter $\sim$30 kpc; thickness of nuclear bulge about 4 kpc; thickness of disk about 700–800 pc; distance between spiral arms about 1.4 kpc. $M_V = -20.5$. Mean density about 0.1 $M\odot$ per cubic parsec. Magnetic field about 3–5 $\times$ 10^{-6} gauss. Total luminosity about 10^{44} ergs s^{-1}.

galaxies, classification of

Hubble's classification of galaxies. Elliptical, ranging from E0 (spherical) to E7 (greatest eccentricity); S0 (nuclei surrounded by disklike structure without arms); spiral, ranging from Sa (arms tightly wound around the nucleus) to Sc (arms widely spread out from the nucleus); barred spirals ranging from SBa (arms tightly wound) to SBc (arms widely spaced out); Irregular (Ir).

Morgan's classification of galaxies. First, the galactic spectral type a, af, f, fg, g, gk, k (corresponding to the integrated stellar types); then the form type S (spiral), B (barred spiral), E (elliptical), I (irregular), Ep (elliptical with dust absorption), D (rotational symmetry without pronounced spiral or elliptical structure), L (low surface brightness), N (small bright nucleus); finally a number from 1 (face-on) to 7 (edge-on). The Andromeda Galaxy (M31) is kS5.

de Vaucouleurs-Sandage classification of spiral galaxies. SA (ordinary spirals), SB (barred spirals); then in parentheses a lower case s (for S-shaped spirals) or r (for the ringed type). Finally, several transitional stages have been added between the SA or SB spirals and the Magellanic irregulars Im. In this classification the Andromeda Galaxy is SA(s)b.

DDO (or van den Bergh) classification of galaxies. This contains two parameters: (1) the galactic type (Sa, Sb, Sc, Ir) and (2) the luminosity class (I, II, III, IV, V), similar to the MKK system of stellar luminosity class. The notations S^- and S^+ are used to denote subgiant species with low and high resolution, respectively. The notation S(B) has been introduced to denote objects intermediate between true spirals and barred spirals.

Radio galaxy optical types. Qs, N, cD, db, d, E.

Galilean satellites   The four largest satellites of Jupiter—Io (J I), Europa (J II), Ganymede (J III), and Callisto (J IV)—discovered by Galileo in 1610. All are locked in synchronous rotation with Jupiter.

gamma   Unit of magnetic field intensity equal to 10^{-5} gauss.

γ The key parameter in special relativity: $\gamma = (1 - v^2/c^2)^{-\frac{1}{2}}$.

gamma-ray bursts   See cosmic gamma-ray bursts.

γ-rays   Photons of very high frequency (wavelengths shorter than a few tenths of an angstrom); the most energetic form of electromagnetic radiation, although there is no discrete cutoff between γ-rays and X-rays. Usually γ-rays come from the nucleus and X-rays come from the inner orbital electrons; however, soft γ-rays and hard X-rays of the same frequency are indistinguishable physically. Galactic γ-rays seem to originate primarily in the spiral arms.

Ganymede (J III)   The largest satellite of Jupiter. Radius 2635 km (slightly larger than Mercury). Mass about 1.65×10^{26} g; period 7.155 days; $e = 0.0015$.

gaseous nebula   An H II region, a supernova remnant, or a planetary nebula. H II regions have an emission-line optical spectrum, and a thermal continuous spectrum declining in intensity as the wavelength increases (from maximum in the ultraviolet) through infrared and radio. Supernova remnants have an emission-line optical spectrum and a nonthermal radio spectrum. Temperatures of planetary nebulae are much higher than those of H II regions.

Gaunt factor (g̅)   A quantum-mechanical correction factor applied to the semiclassical Kramers formula for photon absorption.

gauss   The cgs unit of magnetic flux density. 1 gauss = 1 maxwell per square centimeter = 10^{-4} tesla.

Gaussian distribution   A statistical distribution defined by the equation $p = c \exp(-k^2 x^2)$, in which x is the statistical variable. It yields the familiar bell-shaped curve. Accidental errors of measurement and similar phenomena follow this law.

Gaussian year   The period associated with Kepler's third law with $a = 1$.

gegenschein ("counterglow")   A very faint glow (about 10° across) which can occasionally be seen in a part of the sky opposite the Sun.

α Gem   See Castor.

β Gem   See Pollux.

U Geminorum star   A type of dwarf nova (q.v.). All U Geminorum stars are binaries containing a white dwarf and a red dwarf with total masses of roughly 1–2 $M\odot$ and with periods of less than 12 hours (period of U Gem, 1.5×10^4 seconds). About 150 are known.

YY Gem   See Castor.

general precession   The sum of the lunisolar and the planetary precession (q.v.). It causes the ecliptic longitude to increase at a constant rate (50″.27 per year) but has no effect on ecliptic latitude.

geocorona   The outermost part of Earth's atmosphere, a hydrogen halo extending out to perhaps 15 Earth radii, which emits Lyman-α radiation when it is bombarded by sunlight.

geodetic precession   A small, relativistic, direct motion of the equinox along the ecliptic, amounting to 1″.915 per century.

geoid   The equipotential surface ("mean sea level") of Earth's gravitational field.

geometric albedo   Ratio of the flux received from a planet to that expected from a perfectly reflecting Lambert disk of the same size at the same distance at zero phase angle (cf. Bond albedo).

geometrodynamics (a word coined by John Wheeler)   A theory which attempts to attribute all physical phenomena to properties of spacetime.

giant branch   A conspicuous sequence of red stars with large radii in the H-R diagram of a typical globular cluster that extends from the main-sequence turnoff point upward and redward to the red-giant tip.

gibbous   An adjective applied to the Moon or Venus when it is more than half full (but not full).

giga-   A prefix meaning 10^9.

glitch (Yiddish)   A term used in rocketry to describe a malfunction (or "slide") of the stylus on a chart recorder; also, a sudden change in frequency, as in a pulsar.

globular cluster   A tightly packed, symmetrical group (mass range 10^4–10^6 $M \odot$) of thousands of very old (pure Population II) stars. The stellar density is so great in the center that the nucleus is usually unresolved. Their spectra indicate low abundances of heavy elements. Globular clusters are probably the oldest stellar formation in the Galaxy. They are generally found in the halo and are "high-velocity" objects with very elongated orbits around the galactic center.

globule   A dense, spherical cloud of dust that absorbs radiation (see Bok globule).

Golay cell   A gas bulb used to detect infrared radiation.

Gould Belt   The local system of stars and gas within about 300 pc of the Sun. It is a belt inclined about 10°–20° to the galactic plane in which the greatest concentration of naked-eye O and B stars occurs.

granulation   The mottled appearance of the solar photosphere, caused by gases rising from the interior of the Sun (see granules).

granules   Convective cells (about 1000 km in diameter) in the solar photosphere. Each granule lasts about 5 minutes on the average and represents a temperature roughly 300° higher than the surrounding dark areas. At any one time, granules cover about one-third of the solar photosphere.

gravitation   The universal ability of all material objects to attract each other; $F = Gm_1m_2/r^2$.

gravitational collapse   The sudden collapse of a massive star when the radiation pressure outward is no longer sufficient to balance the gravitational pressure inward. In gravitational collapse there is a sudden, catastrophic release of great quantities of gravitational potential energy, and this release has been postulated as the cause of supernovae, neutron stars, and black holes.

gravitational constant (*G*)   The constant of proportionality in the attraction between two unit masses a unit distance apart. $G =$ 6.668 $\times$ 10^{-8} dyn cm^2 g^{-2}.

gravitational encounter   The encounter between two massive bodies which results in the deviation from their original directions of motion.

gravitational equilibrium   The condition in a star in which at each point the weight of the overlying layers is balanced by the total pressure.

gravitational-lens effect   The effect of matter in curved spacetime, which tends to focus any beam of radiation from a distant source. In effect, the spacetime curvature is a lens of great focal length. At $z \approx 1$, the angular size of an object starts increasing with distance.

gravitational mass   That property of matter which makes it create a gravitational field and attract other particles (cf. inertial mass; equivalence principle).

gravitational radiation   According to general relativity, any massive body with variable quadrupole and higher moments must emit gravitational waves. In the lowest approximation, the radiation emitted is derived from the varying quadrupole moment of the object; more generally, any body which experiences varying acceleration will emit gravitational waves by an amount proportional to the rate of change of the acceleration.

gravitational radius   The radius which an object should have in order that light emitted from its surface just ceases to escape from its surface.

gravitational redshift   The rate at which a clock keeps time when it is in a gravitational field is slower than the rate at which it will keep time in the absence of a gravitational field. (The gravitational redshift was experimentally verified by Pound and Rebka in 1960.) The amount of redshift is directly proportional to the mass of the emitting body and inversely proportional to its radius.

graviton   A hypothetical elementary particle associated with the gravitational interaction. It is a stable particle with zero rest mass, zero charge, and a spin of ± 2, and travels with the speed of light.

gravity darkening   See von Zeipel's theorem.

gray atmosphere   A model atmosphere in which the continuous absorption coefficient is assumed to be independent of frequency.

gray body   A body whose emissivity is constant and less than unity.

grazing-incidence telescope   A telescope used in X-ray and gamma-ray astronomy. It focuses these rays by making use of the fact that they behave like light rays if they strike surfaces at a shallow enough angle.

Great Looped Nebula   See 30 Doradus.

Great Red Spot   See Red Spot.

Great Rift   A "split" in the Milky Way between Cygnus and Sagittarius caused by a succession of large, overlapping dark clouds in the equatorial plane of the Galaxy. It is about 100 pc distant.

Green's theorem   An identity that connects line integrals and double integrals.

Grotrian diagram   Energy-level diagram.

ground state (of an atom)   The state in which all electrons are in the lowest possible energy states.

guillotine factor   A factor that measures the sharp reduction in the opacity of a gas when the temperature is high enough to have ionized the atoms down to their K shells.

Gum Nebula   A giant H II region 30°–40° in diameter in which the Vela pulsar and the Vela X supernova remnant are embedded. It appears to be a fossil Strömgren sphere produced by an outburst of ionizing radiation that accompanied the Vela X supernova remnant. (Named for the Australian astronomer Colin Gum.)

gyrofrequency   The frequency with which an electron or other charged particle executes spiral gyrations in moving across a magnetic field.

gyrosynchrotron radiation   Radiation emitted by *mildly* relativistic electrons.

H

$\hbar$ The quantity h (Planck's constant) divided by 2π: $\hbar = 1.054 \times 10^{-27}$erg seconds.

h Hubble's constant in units of 100 km s^{-1} Mpc^{-1}.

h-line   An Mg II resonance line at 2803 Å.

H and K lines   The two closely spaced lines of singly ionized calcium at 3968 and 3934 Å, respectively.

H$^-$ ion   An H ion with an extra electron in its shell. It is an important source of stellar opacity in stars whose spectral types are later than about A5.

H-magnitude   The magnitude derived from infrared observations at 1.6 microns.

H I region   Region of neutral (atomic) hydrogen in interstellar space. The temperature is about 125 K (the spin temperature of neutral hydrogen)—far too low for electrons to emit radiation in the optical part of the spectrum (see 21-cm radiation). At least 95 percent of interstellar H is H I. (Density is about 10 atoms per cm^3, about the same as in H II regions.)

H II condensation   A high-density H II region.

H II region   Region of ionized hydrogen in interstellar space. H II regions occur near stars with high luminosities and high surface temperatures. The kinetic temperature of H II regions is about 10,000–20,000 K, and the density is about 10 atoms per cm^3. Ionized hydrogen, of course, having no electron, does not produce spectral lines; however, occasionally a free electron will be captured by a free proton and the resulting radiation can be studied optically (see also radio recombination lines).

H-R diagram   See Hertzsprung-Russell diagram.

Hades   An unofficial name for Jupiter IX, the outermost satellite of Jupiter ($P = 758$ days retrograde, $i = 156°$, $e = 0.28$). Discovered by Nicholson in 1914.

hadron   Any elementary particle (baryon or meson) that interacts strongly in nuclei.

hadron barrier   The interval ($t \approx 10^{-43}$ [10^{-23}] s after the big bang, when $\rho \approx 10^{93}$ [10^{52}] g cm^{-3}) during which quantum and general-relativistic effects are expected to modify each other in an unknown way. The quantities in brackets are for a different equation of state.

hadron era   The interval ($t \approx 10^{-5}$ s after the big bang) when the Universe was matter-dominated and when $kT \approx m_\pi$. It was followed by the lepton era (q.v.).

Hagedorn equation of state   An equation of state for extremely degenerate matter (density greater than about 10^{15} g cm^{-3}).

halation   The formation of a halo around bright star images by light reflected from the back of the photographic plate or emulsion.

half-life   For any radioactive substance, the length of time required for half the atoms to disintegrate (cf. mean life).

half-power beamwidth (HPBW)   The angle across the main lobe of an antenna pattern between the two directions where the sensitivity of the antenna is half the value at the center of the lobe. This is the nominal resolving power of the antenna system.

Halley's comet   Probably the best known of all comets. Its orbit was computed by Edmund Halley in 1704, at which time he predicted that the bright comet of 1682 would return in 1758 (Halley died in 1742, before he could see his prediction verified). Records of Halley's comet ($a = 17.8$ AU, $e = 0.967$, $i = 162°.3$, $P = 76.2$ yr, perihelion distance 0.587 AU) have been traced back to 240 B.C. Last appearance 1910, next appearance 1986.

halo (of galaxy)   The tenuous, spherical cloud surrounding a spiral galaxy. It is the locus of old stars and globular clusters. Our Galaxy and many others are surrounded by a halo of very hot gas emitting in the X-ray region. These halos contain relativistic electrons which, when deflected by the magnetic field, emit radiation in the radio band. The halo of our galaxy has a radius of about 15 kpc.

halo population   Old stars typical of those found in the halo of the Galaxy; also called Population II.

Hamiltonian function (H)   The quantity in classical mechanics corresponding to the total energy of a system, expressed in terms of momenta and positional coordinates.

Hamiltonian operator (H)   The dynamical operator in quantum mechanics that corresponds to the Hamiltonian function in classical mechanics.

Hanning method   A method of smoothing out the noise in radio data. For each data point, one-half the value of that point is taken, plus one-quarter the value of the point on each side. The result is usually a smoother curve.

Harkins's rule   The rule that atoms of even atomic number are more abundant than those of odd atomic number.

Harman-Seaton sequence   An evolutionary sequence of hot subdwarfs and nuclei of planetary nebulae.

harmonic law   See Kepler's third law.

harmonic motion (also called periodic motion)   A motion that repeats itself in equal intervals of time.

harmonic oscillator   Any oscillating particle in harmonic motion.

harmonic overtone   Any integral multiple of the fundamental frequency (q.v.).

Haro galaxies   Blue objects whose spectra show sharp emission lines.

Harvard classification   See Henry Draper system.

Hawking's theorem   (1) A stationary black hole must be either static (i.e., nonrotating) or axisymmetric. (2) In interactions involving black holes, the surface area of the event horizon can never decrease.

Hayashi track   A nearly vertical track of stellar evolution toward the main sequence during phases when the star is largely or completely in convective equilibrium. The luminosity, originally very high, decreases rapidly with contraction, but the surface temperature remains almost constant.

head (of comet)   See coma.

"head-tail" galaxies   A class of relatively weak radio sources associated with clusters of galaxies and characterized by a high-brightness "head" close to the optical galaxy and a long low-brightness "tail."

Heaviside layer   See E layer.

heavy-metal stars   A class of peculiar giants that includes the Ba II stars and the S stars.

Hektor   Asteroid 624, the largest (about 100 km long) of the Trojans. Its shape is apparently as elongated as that of Eros. Rotation period 6.9225 hours. Its visual mean opposition magnitude is near +14.5, which makes it the brightest of the Trojans. Assumed albedo 0.28. It has a large obliquity.

helium flash   The onset of runaway helium burning under degenerate conditions. The helium flash occurs in the hydrogen-exhausted core of a star in the red-giant phase of evolution. When gravitational pressure has brought the degenerate core to a temperature of about 10^8 K, the helium nuclei can start to undergo thermonuclear reactions. Once the helium burning has started, the temperature builds up rapidly (without a cooling, stabilizing expansion), and the extreme sensitivity of the nuclear reaction rate to temperature causes the helium-burning process to accelerate. This in turn raises the temperature, which further accelerates the helium burning, until a point is reached where the thermal pressure expands the core and thus removes the degeneracy and limits the flash. The helium flash can only occur when the helium core is less than the $1.4M\odot$ Chandrasekhar mass limit and thus it is restricted to low-mass stars.

helium shell flash   It has been shown that helium shell burning outside a degenerate core is unstable; the helium-burning shell does not generate energy at a constant rate, but instead produces energy primarily during short flashes. During a flash, the region just outside the helium-burning shell becomes unstable to convection. The resultant mixing probably leads to the *s*-process as well as to the upward movement of carbon produced by helium burning.

Helix Nebula (NGC 7293)   A planetary nebula about 140 pc distant in Aquarius with the largest known angular diameter of any planetary.

Helmholtz contraction   See Kelvin-Helmholtz contraction.

Henry Draper system   A classification of stellar spectra into the sequence O, B, A, F, G, K, M, in order of decreasing temperature.

Henyey track   An almost horizontal track of stellar evolution between the Hayashi track and the main sequence.

Hera   Unofficial name for Jupiter VII. $P = 259.65$ days, $e = 0.21$, $i = 28°$. Discovered by Perrine in 1905.

Herbig-Haro object   An object with many of the characteristics of a T Tauri star (e.g., its spectrum shows a weak continuum with strong emission lines), believed to be a star in the very early stages of evolution. All known Herbig-Haro objects have been found within the boundaries of dark clouds. They are strong infrared sources and are characterized by mass loss.

Hercules cluster (3U 1551 + 15)   An unsymmetrical cluster of about 75 bright galaxies ($z = 0.036$) of which about half are spiral or irregular and about half elliptical or S0. It contains a rather large number of disturbed and peculiar galaxies. The "missing mass," if present, must constitute more than 95% of the total.

Hercules X-1 (3U 1653 + 35)   An X-ray pulsar probably about 5 kpc distant, a member of an occulting binary system with an orbital period of 1.7 days. The visible component has been identified as the blue variable HZ Herculis, whose spectrum varies from late A or early F to B. Her X-1 has a pulsation period of 1.2378 seconds, presumably its rotation period, and exhibits a 35-day quasi-periodicity in the X-ray region (but not in the optical). It is probably a rotating neutron star in a circular orbit ($e < 0.1$) with a mass of about $0.7\ M\odot$, which is accreting matter from HZ Her. The orbital period is stable, but the pulsation period is speeding up at a rate of about 1 part in 10^5 per year. The X-ray eclipse lasts 0.24 days.

DQ Herculis = Nova Herculis 1934   A slow nova. It is an eclipsing binary with an orbital period of only 4^h39^m. It also has a regular flickering period of 71 seconds, the shortest period of regular variations known, except for pulsars and compact X-ray objects. It is probably composed of an M dwarf and a white dwarf with an accretion disk.

Hermitian matrix   A matrix which remains unchanged if each element is replaced by its complex conjugate and the rows and columns are interchanged. In quantum mechanics all matrices corresponding to observables have this property.

hertz (Hz)   A unit of frequency equal to 1 cycle per second.

Hertzsprung gap   A gap (from about A0 to F5) in the horizontal branch of the H-R diagram (see instability strip). The few stars that populate this gap are RR Lyrae and other variable stars. It is regarded as a region through which a star moves rapidly in its evolutionary track away from the main sequence.

Hertzsprung-Russell diagram (H-R diagram)   In present usage, a plot of bolometric absolute magnitude against effective temperature for a group of stars. Related plots are the color-magnitude plot (absolute or apparent visual magnitude against color index) and the spectrum-magnitude plot (visual magnitude versus spectral type)—the original form of the H-R diagram.

Hess diagram   A diagram showing the frequencies with which stars occur at various positions in an H-R diagram.

Hestia   Unofficial name for Jupiter VI. Discovered by Perrine in 1904. $P = 250$ days, $e = 0.16$, $i = 29°$.

Hidalgo   Asteroid 944, perhaps 20 km in diameter, with the largest known orbit ($a = 5.8$ AU), second highest inclination to the ecliptic ($42°.5$), and second highest eccentricity ($e = 0.66$) of any known minor planet. Period 13.7 years. Discovered by Baade in 1920.

"hidden mass"   See mass discrepancy.

hierarchical cosmology   A cosmology characterized by a system of clusters within clusters within clusters.

"high-velocity" object   Generally a celestial object in the galactic halo whose orbital velocity around the galactic center is less than that of the Sun, and that thus, relative to the Sun, has a high space motion. A "high-velocity" object usually travels around the galactic center in an eccentric orbit, often of large inclination to the galactic plane.

Hind's nebula (NGC 1554-5)   A reflection nebula (q.v.) discovered by Hind in 1852, which is illuminated by the star T Tauri. It is remarkable for its changes in brightness.

Hirayama families   Groups of minor planets with similar orbital elements. The members of a given family are widely believed to have resulted from collisions between larger parent bodies.

Holmberg radius   The radius of an external galaxy at which the surface brightness is 26.6 mag arcsec^{-2}. This criterion was developed by Holmberg in 1958 to estimate the actual dimensions of the major and minor axes of a galaxy without regard to its orientation in space.

Holtsmark approximation   An approximation in which the lines emitted and absorbed by atoms are subject to the fluctuating electrostatic fields to which the atom is subject in an ionized atmosphere.

horizontal branch   That part of the H-R diagram of a typical globular cluster that extends shortward from the asymptotic branch at an approximately constant absolute bolometric magnitude of about 0.3. A star appears on the horizontal branch after it has undergone the helium flash and begins to burn helium quietly in its core and hydrogen in a surrounding envelope.

Horsehead Nebula (NGC 2024)   An absorption nebula in the middle of Orion.

Horseshoe Nebula   See Omega Nebula.

hour angle   The angle (measured westward) between the meridian and an hour circle. The hour angle depends both on time and on the observer's location. It can be determined by subtracting the star's right ascension from the sidereal time.

hour circle   A great circle passing through the celestial poles—i.e., perpendicular to the celestial equator.

Hourglass Nebula   A compact H II region in the center of M8.

Hoyle-Narlikar theory   A reformulation of the general theory of relativity that incorporates and extends Mach's principle (q.v.). In this theory, the inertial mass of a particle is a function of the masses of all other particles, multiplied by a coupling constant which is a function of cosmic epoch. In cosmologies based on this theory, the gravitational constant G decreases strongly with time.

Hubble constant (H_0)   The constant of proportionality in the relation between the recession velocities of galaxies and their distances from us. The value determined by Sandage and Tammann (1974) is 56.9 ± 3.4 km per second per megaparsec.

Hubble diagram   Plot of apparent magnitude of galaxies versus their redshift.

Hubble law (also called law of redshifts)   The distance of galaxies from us is linearly related to their redshift. More generally, the relative velocity between particles that are locally at rest is proportional to their separation.

Hubble nebula (NGC 2261)   A cometary nebula whose apex star is R Mon.

Hubble radius (c/H)   The radius of the observable universe ($> 10^{27}$ cm).

Hubble time (H_0^{-1})   The assumed age of the Universe since the big bang (17.7×10^9 yr for $H_0 = 55$ km per second per megaparsec and a constant expansion rate).

Hugoniot relations   Relations expressing conservation of baryon number, momentum, and energy across a shock front.

Hulse-Taylor pulsar (PSR 1913 + 16)   A binary pulsar discovered in 1974, probably consisting of a neutron star and an even more compact object in an eccentric orbit, with an orbital period of 0.3230 days and a pulsation period of 59 milliseconds.

Hund's rule   The larger the value of S (total spin angular momentum), the lower the value of the average perturbation energy $\langle V \rangle_{SL}$.

Huyghenian region   The brightest portion of the Orion Nebula.

Hyad   A single member of the Hyades.

Hyades   A young (5×10^8 yr) moving cluster (radial velocity, $+36$ km s^{-1}) of more than 200 stars (spectral types A1–K) visible to the naked eye in Taurus, about 40 pc distant. Aldebaran is a foreground star in that region of the sky.

hydromagnetics   See magnetohydrodynamics.

hydrostatic equilibrium   A balance between the gravitational force inward and the gas and radiation forces outward in a star.

hypercharge   Twice the charge of a charge multiplet (q.v.).

hyperfine structure   Splitting of spectral lines due to the spin and consequent magnetic moment of an atomic nucleus. It can be observed only at very high resolution.

Hyperion   Eighth satellite of Saturn about 160 km in diameter. $P = 21^d6^h38^m$. Discovered by Bond in 1848.

hyperon   An unstable heavy baryon (q.v.) with an average life of 10^{-8} to 10^{-10} seconds.

HZ stars   Blue horizontal-branch stars, the first catalog of which was compiled by Humason and Zwicky.

I

Iapetus   The ninth satellite of Saturn, about 850 ± 100 km in radius; period $79^d7^h55^m$, $e = 0.028$, inclination to Saturn's orbital plane $14°.7$. It has the most extreme variation in albedo of any satellite in the solar system (0.04 for the leading side, 0.28 for the trailing side). Discovered by Cassini in 1671.

Icarus   Asteroid No. 1566, 1.1 km in diameter, discovered by Baade in 1948. It has the smallest orbit and highest eccentricity ($a = 1.07$ AU, $e = 0.827$, $i = 23°$, $P = 408^d$) of any known minor planet. It is the only asteroid known to come closer to the Sun than Mercury (perihelion distance 0.19 AU). Rotation period 2^h16^m.

ideal gas (also called perfect gas)   A nondegenerate gas in which the individual molecules are assumed to occupy mathematical points and to have zero volume, and in which the mutual attraction of neighboring molecules is zero.

ideal gas laws   The pressure of a gas is directly proportional to the product of its temperature and density ($p = C\rho T$). The higher the temperature and the more rarefied a gas, the more closely it obeys the ideal gas laws, so the gases in most stars closely approximate ideal gases. For a degenerate gas, the pressure depends only on the density and is independent of the temperature.

image tube (also called image intensifier)   An electronic camera in which electrons, emitted from a photocathode surface exposed to light, are focused electronically onto a phosphor or photographic plate.

immersion   The disappearance of a celestial body due to eclipse or occultation.

impact parameter   A measure of the distance by which a collision misses being head-on. In astronomy, usually the distance, at closest approach, between the centers of two particles in a collision if there were no attractive force acting between them.

imperfect scattering   See nonconservative scattering.

inclination   The angle between the plane of a planet's or satellite's orbit and the plane of the ecliptic (cf. obliquity), or between the plane of a satellite's orbit and the planet's equatorial plane.

index of refraction (n)   The ratio of the speed of light in a vacuum to that in a given medium.

induced emission   See stimulated emission.

inelastic collision   A reaction involving a change in the kinetic energy of the system, as in ionization, excitation, or capture; or a process which changes the energy level of the system.

inertial frame of reference   Any "standard of rest" or coordinate frame for which Newton's first law is valid.

inertial mass   The property of matter which gives it inertia (cf. gravitational mass).

infrared   That part of the electromagnetic spectrum that lies beyond the red, having wavelengths from about 7500 Å to a few millimeters (about 10^{11}–10^{14} Hz). Infrared radiation is caused by atomic transitions, or by vibrational (near-IR) and rotational (far-IR) transitions in molecules. (IRe1, IRc1, IRs1: the e sources are extended; the c sources are unresolved; the s indicates an infrared nebula surrounding a visible star.)

initial mass function (IMF)   A quantity that determines the number of stars per unit time evolving from the main sequence.

inner bremsstrahlung   The continuous electromagnetic radiation that accompanies the β-decay of nuclei.

inner Lagrangian point (L_1)   The Lagrangian point (q.v.) through which mass transfer occurs.

insolation (the word is a contraction of "incoming solar radiation")   Amount of radiation received from the Sun per unit area on the Earth's surface per unit time.

instability strip   A region in the Hertzsprung gap (q.v.) occupied by pulsating stars in a post–main-sequence stage of stellar evolution. Stars traverse the instability strip relatively quickly at least once on their way to their final evolutionary configuration.

intensity (sometimes called specific intensity)   A measure of the amount of light received per unit solid angle per unit time per unit area normally from an element of surface.

interactions   There are four known kinds of interactions between particles. In the order of decreasing strength they are the strong interactions (effective only to distances of about 3 fermis); electromagnetic interactions, which are the interactions of charged particles with electromagnetic fields; weak interactions, responsible for β-decay; and gravitational interactions, the weakest of all. The pion is associated with the strong interactions; the photon, with electromagnetic interactions; the neutrino, with weak interactions; and the graviton, with gravitational interactions. If gravitational $= 1$, then weak $= 10^{28}$, electromagnetic $= 10^{39}$, strong $= 10^{41}$.

intercloud medium   The sparsely populated (about 0.5 atoms per cm^3) regions of space between the interstellar concentrations of gas and dust.

intercombination lines   Spectral lines emitted in transitions between two levels with different values of S.

interference   Alternate reinforcement and cancellation of two or more beams of electromagnetic radiation from the same source. In constructive interference the two component beams are in phase, and light results. In destructive interference the components are out of phase, and darkness results.

interference filter   A filter used to shut out all light except the desired wavelengths.

interferometer (stellar)   A device for measuring small angles by using the principle of interference (q.v.). In radio astronomy, it consists of two or more separate antennas at some distance from one another and each receiving radiation from the same source, joined to the same receiver. The advantage of interferometers is that they reject background noise; their disadvantage is that they are sensitive only to radio waves from sources of small angular size.

intergalactic medium   Hypothetical matter (in the form of gas) in the regions between galaxies. It has not yet been detected (but see Magellanic Stream), but Oort (1970) has argued that as much as a factor of 16 or more matter may be present in uncondensed form.

intermediate vector boson (IVB) (sometimes called W boson)   A hypothetical elementary particle that acts as intermediary for the weak interaction, carrying its effect from one particle to another as the photon does for electromagnetic interactions and as various mesons do for the strong interactions.

International System of Units (SI units)   A practical system of units of measurement adopted in 1969 by the 11th International General Conference of Weights and Measures (CGPM). The seven base units are the meter, the kilogram, the second, the ampere, the kelvin, the mole, and the candela.

interpulse   The weaker component of a pulsar pulse when its period is roughly half that of the main pulse.

interstellar dust   Small grains (primarily silicates) in the interstellar gas. Their dimension is roughly that of optical wavelengths—i.e., about 4000–7000 Å—so they absorb photons of visible light and reemit in the far-infrared. The particles are polarized, needle-shaped grains. Interstellar dust affects the entire spectrum, and leads to dimming and reddening of starlight. Temperature of interstellar dust 5–20 K.

interstellar extinction   The reddening of starlight passing through interstellar dust, caused by the fact that dust scatters blue light more than red.

interstellar gas   Sparse, cool gas (mainly hydrogen) in interstellar space. Dust absorbs and scatters radiation; gas does not interact directly with radiation but is coupled to the dust by collisions. Interstellar gas affects only light of certain wavelengths. Temperature 10–100 K.

interstellar grains   See interstellar dust.

interstellar lines   Sharp, distinct absorption lines superposed on stellar spectra, produced by the interstellar gas located between the source and the observer. Strongest are the D lines, followed by the H and K lines, and the K I doublet at 7699 and 7644 Å.

interstellar matter   Interstellar gas (99%) and dust (1%). The observed density of the interstellar medium is about 1–5 atoms per cm³. The two other components of the interstellar medium are magnetic fields and cosmic-ray electrons.

interstellar medium   See interstellar matter.

interstellar molecules   Molecules in interstellar space. As of late 1974, at least 33 molecular species had been identified with reasonable certainty: methylidyne CH, ionized methylidyne CH⁺, the cyanogen radical CN, the hydroxyl radical OH, ammonia NH_3, water vapor H_2O, formaldehyde H_2CO, carbon monoxide CO, hydrogen cyanide HCN, hydrogen isocyanide HNC, molecular hydrogen H_2, X-ogen, cyanoacetylene HC_3N, methyl alcohol CH_3OH, formic acid HCOOH, carbon monosulfide CS, carbonyl sulfide OCS, formamide NH_2CHO, silicon monoxide SiO, methyl cyanide CH_3CN, isocyanic acid HNCO, methyl formate $HCOOCH_3$, methyl acetylene CH_3C_2H, acetaldehyde CH_3CHO, thioformaldehyde H_2CS, hydrogen sulfide H_2S, methanimine H_2CNH, ethynyl, sulfur monoxide SO, dimethyl ether $(CH_3)_2O$, methyl amine CH_3NH_2, silicon monosulfide SiS, and ethyl alcohol C_2H_5OH.

interstellar reddening   See interstellar extinction.

interval   The "distance" between two events in four-dimensional spacetime.

invariant   An adjective referring to a quantity whose numerical value is the same in all coordinate systems.

invariant plane (solar system)   The plane defined by the total angular momentum of the solar system. It is within about 1.̊5 of the ecliptic.

inverse β-decay   The relatively rare process $p + \bar{\nu} \to n + e^+$. Free-electron capture ($e + p \to n + \nu$) is sometimes called inverse β-decay in astrophysics.

inverse bremsstrahlung   Absorption (free-free absorption) of a photon by an electron in the field of a nucleus.

inverse Compton effect   The collision between a photon and an energetic (cosmic-ray) electron, in which some of the energy of the electron is transferred to the photon.

inverse maser   A mechanism that absorbs radiation and cools a gas so that the number of molecules in an upper level falls below that expected in a condition of thermal equilibrium. The effect is inverse to that found in a maser, where there is an overpopulated upper level.

inverse P Cygni profile   A profile in which the emission is on the violet side of the absorption. It is usually interpreted to mean infall of matter.

inverse plasmon scattering   Scattering of electrostatic plasma waves by a flux of relativistic electrons.

Io (Jupiter I)   The innermost Galilean satellite of Jupiter, similar in size and density to the Moon ($R \approx 1850$ km from *Pioneer 10*, period 1.77 days; $e \approx 0.01$, $i = 0.03$). Jupiter's decametric radiation has been linked at least partially to Io. Mean density (from *Pioneer 10*) 3.48 g (the highest of any of the Galilean satellites). *Pioneer 10* also detected the presence of an ionosphere, and Na D emission. Albedo 0.91(?), the highest in the solar system.

ionization potential   The minimum energy required to remove an electron from an atom. It always takes a higher energy to remove a second electron from a singly ionized atom, a still higher energy to remove a third, etc. The ionization potential for hydrogen is 13.596 eV, which corresponds to a wavelength of 912 Å.

ionosphere   The region of Earth's atmosphere (80–500 km), immediately above the stratosphere. The ionosphere consists of the D layer, the E layer, and the F layers (q.v.). It is strongest at the end of the day.

iron peak   A maximum on the element-abundance curve near atomic mass number 56.

irregular galaxy   A galaxy with amorphous structure and with relatively low mass (10^8–10^{10} $M\odot$). Fewer than 10% of all galaxies are classified as irregular.

isobars   Nuclei with the same A number but different Z numbers. Also, lines connecting equal atmospheric pressures.

isochoric   Constant volume.

isochrones   Time-constant loci.

isoelectronic sequence   A sequence of ions which have the same number of electrons but different atomic numbers.

isomers   Nuclei with the same A and Z numbers but in different energy states.

isomer shift   Displacement of an absorption line due to the fact that the absorbing nuclei have a different s-electron density from that of the emitting nuclei.

isophotes   Lines connecting points of equal light intensity.

isospin (also called isotopic spin)   A quantum number which arises from regarding different members of a charge multiplet (q.v.) as different states of a single particle.

isostasy   The plasticity of the surface layer of a planet—e.g., the ability of the surface layer to adjust its level according to the load (such as ice caps) that it has to carry.

isotones   Nuclei with the same number of neutrons but with different A and Z numbers.

isotopes   Nuclei with the same Z number but with different A numbers.

isotopic spin   See isospin.

Israel-Robinson theorem   The only locally stationary empty space-time which is asymptotically flat with a nondegenerate event horizon is the $|a| < m$ Kerr metric, where a is angular momentum per unit mass.

J

J-file   A group of lines of a supermultiplet having a common lower level.

j-j coupling   See LS coupling.

J-magnitude   The magnitude derived from the observations at an infrared wavelength of 1.3 microns.

J-value   Value of the total angular momentum (orbital plus spin). J is the rotational quantum number which specifies the rotational level of a molecule.

Jacobi ellipsoid   Jacobi discovered that homogeneous, self-gravitating masses rotating uniformly and sufficiently rapidly can have the shape of triaxial ellipsoids. These are the Jacobi ellipsoids.

jansky (Jy)   Unit of flux density adopted by the IAU in 1973. 1 Jy = 10^{-26} W m^{-2} Hz^{-1}.

Janus   The innermost satellite of Saturn, just outside Saturn's rings. $P = 0.75$ days; $R = 175$ (?) km; $i \approx 0$; $e \approx 0$. It was discovered by Dollfus in 1966 and was named Janus for the first and the last.

Jeans instability criterion   (See Jeans length)

Jeans length   The critical wavelength ($\lambda_J = c_s(\pi/G\rho_0)^{1/2}$, where c_s is the isothermal sound speed in the medium) at which the oscillations in an infinite, homogeneous medium become gravitationally unstable. Any disturbance greater than the Jeans length will decouple by self-gravitation from the rest of the medium to become a stable, bound system. In general, $\lambda_J \approx 10^{20}$ cm.

jitter   Irregular random variations in a radio signal.

Johnson-Morgan system   See UBV system.

Johnson noise   Low-frequency electromagnetic radiation associated with thermal fluctuations, which is emitted by all bodies whose temperature is above 0 K.

joule   The SI unit of energy, work, or quantity of heat. 1 J is equal to a force of 1 newton acting over a distance of 1 meter. $1 J = 10^7$ ergs.

Joule dissipation   The heat produced when a current is passed through an electrically resisting medium.

Julian date (JD)   The number of ephemeris days that have elapsed since 12^h ephemeris time on January 1, 4713 B.C. JD for 1970 January 1 is 2440588.

jump conditions (in a shock wave)   The conditions for jumps in pressure and density (or temperature or energy) across the shock. These are the Hugoniot conditions.

Juno   An asteroid 250 km in diameter ($P = 1,594$ days; $a = 2.67$ AU; $e = 0.256$; $i = 13°.0$) with a relatively large albedo (0.2). Rotation period 7^h21.

Jupiter   Fifth planet from the Sun. Mass 1.90×10^{30} g $= 318$ times Earth's. It is more massive than all other planets and satellites combined; if it were about 80 times more massive, it would become self-luminous. Equatorial radius 7.135×10^9 cm $= 11$ times Earth's; polar radius 6.7×10^9 cm; oblateness (from *Pioneer 10*) 0.065. Surface gravity 2.7 that of Earth; V_{esc} 61 km s^{-1}; mean density 1.33 g cm^{-3}. Rotational period 9^h50^m at equator; 9^h55^m at polar regions (see Systems I and II longitude). Semimajor axis 5.203 AU, $e = 0.048$, $i = 1°18'18''$. Obliquity 3°.1. Orbital period 11.86 years, mean orbital velocity 13.06 km s^{-1}. Synodic period 398.9 days. Albedo 0.51. Surface temperature about 120 K. Current estimates of Jupiter's central temperature and density are 5×10^4 K and about 35 megabars (the heat flux to the surface is mainly convective). For both Jupiter and Saturn it is necessary to invoke a substantial source of internal heating (presumably gravitational contraction) to account for the surface temperature (Jupiter radiates about 2½ times as much heat as it receives from the Sun). Jupiter's core is probably metallic hydrogen in a pressure-ionized liquid phase. Jupiter's surface shows pronounced horizontal striations: the light layers (zones) are at a slightly higher altitude and about 15° cooler than the dark layers (belts). Atmosphere primarily H_2 (85%) and He (14%), with traces of methane, ammonia, etc. *Pioneer 10* established that Jupiter has a magnetic field of about 4 gauss (the magnetic axis is inclined 15° to the rotational axis and is offset about 0.1 Jupiter radius from the center of the planet), and also that it is surrounded by a partial torus of atomic H in the orbit of Io. Thirteen satellites, the four outermost of which have retrograde motion, high eccentricity, and high inclination. (Jupiter XIII, discovered in 1974, has a period of 239 days; $i = 26°.7, e = 0.147$.)

K

K   A quantum number which refers to the component of angular momentum around a molecule's axis of symmetry.

Kα   A spectral line in the X-ray region ($\lambda = 0.334$ Å), produced by the transition between the lowest level of the K shell and the lowest level of the L shell.

K-capture   Capture by a nucleus of one of the electrons in its innermost shell, accompanied by the emission of X-rays.

K component (from the German *Kontinuum*)   The inner part of the solar corona (the gaseous phase) which emits a continuous spectrum without absorption lines. Physically, the K component results from Thomson scattering of photospheric radiation by free electrons in the corona. The K component is polarized and decreases rapidly with distance from the Sun.

K corona   See K component.

K-correction (also called *K*-term)   A correction term that must be applied to all photographically observed magnitudes of distant galaxies to allow for the alteration of the spectrum due to the redshift and to absorption by dust.

K-edge   The absorption edge of the K shell (see absorption edges).

K-electron   An electron in the K shell.

k line   A Mg II resonance line at 2795.5 Å.

K line   A spectral line of singly ionized calcium at 3933 Å (see Fraunhofer lines).

K-magnitude   The magnitude derived from observations at 2.2 microns.

K shell   The innermost shell, or energy level, of an atom. All elements heavier than hydrogen have a filled K shell, which consists of two $1s$ electrons orbiting the nucleus.

K star   Stars of spectral type K are cool, orange to red stars with surface temperatures of about 3600–5000 K. Their spectra resemble those of sunspots, in which the hydrogen lines have been greatly weakened. The H and K lines (q.v.) reach their greatest intensity. Strongest lines are Ca I (4227 Å) and the G-band (4303 Å). Examples are Arcturus and Aldebaran.

K-term   See *K*-correction.

Kapteyn's star (HD 33793, CD − 45° 1841)   A high-velocity (radial velocity + 242 km s⁻¹) M0 subdwarf 3.9 pc distant.

kayser (= cm⁻¹)   A wavenumber.

Keller-Meyerott opacity   Opacity taken from the tables derived by G. Keller and R. Meyerott for assumed mixtures of hydrogen, helium, and heavy elements ranging from $X = 99\%$, $Y = 1\%$, $Z = 0$ to $X = 0$, $Y = 50\%$, $Z = 50\%$.

kelvin (K)   The SI unit of thermodynamic temperature. "The kelvin, unit of thermodynamic temperature, is the fraction 1/273.16 of the thermodynamic temperature of the triple point of water." [13th CGPM (1967), Resolution 4]

Kelvin contraction (also called Kelvin-Helmholtz contraction)   The contraction of a star contemplated by Kelvin and Helmholtz as a consequence of a star's radiating its thermal energy. It is currently believed that the contraction of a star occurs in this manner in its pre-main-sequence evolution.

Kelvin scale   A temperature scale with the same divisions as the Celsius (centigrade) scale and with the zero point at 0° absolute. (Room temperature is about 295 K).

Kelvin time scale   The time it takes a star to contract gravitationally from infinite radius down to its present radius by radiating its thermal energy (for the Sun, about 2–3 × 10⁷ years). The Kelvin time is roughly equal to (gravitational binding energy)/luminosity. (Compare nuclear time scale).

Kennelly-Heaviside layer   Former name for the D and E layers (q.v.).

Kepler's laws   (1) Each planetary orbit is an ellipse with the Sun at one focus. (2) (law of areas) Equal areas are swept out in equal times. (3) (harmonic law) The square of the period is proportional to the cube of the distance. Newton's generalized formula for the third law is $P^2 = 4\pi^2 a^3/[G(m_1 + m_2)]$.

Keplerian orbit   The orbit of a spherical particle of a finite mass around another spherical particle, also of finite mass, by virtue of the gravitational attraction between them. In the Bohr-Sommerfeld picture of atoms, the electrons are considered as describing Keplerian orbits in the field of the positive nucleus by virtue of the inverse square electric attraction between the electrons and the nuclei.

Kepler's supernova (3C 358)   A Type I supernova (SN Oph 1604) whose light reached Earth in 1604. If $H_0 = 50$, then Kepler's

supernova is out in the galactic halo at a distance of 12.1 kpc
and 1.4 kpc above the galactic plane, according to van den
Bergh. Kepler's supernova is the prototype of Type I supernovae; at its brightest it reached an apparent magnitude of about
−2.2.

kernel   The set of points mapped into zero.

Kerr black hole   A rotating, axisymmetric black hole based on Kerr's
1963 solution to Einstein's field equations.

Keyhole Nebula   An old name for Eta Carinae.

Killing vector   When it exists, the Killing vector describes symmetry
properties of spacetime. If spacetime admits a timelike Killing
vector, then it is stationary; similarly, axial symmetry is derived
from a spacelike Killing vector.

kilo-   A prefix meaning 10^3.

kilogram   The SI basic unit of mass (not of weight or of force). 1
kilogram is equal to the mass of 1.000028 cubic decimeters of
water at the temperature of its maximum density.

kinematics   The branch of mechanics that studies bodies undergoing
continuous change of position. Whereas dynamics takes into account mass, force, distance, and time, kinematics is concerned
only with distance and time.

kinetic temperature   A measure of the average random motion of the
particles in a system.

Kirchhoff's laws   (1) To each chemical species there corresponds a
characteristic spectrum. (2) Every element is capable of absorbing the radiation which it is able to emit; this is the phenomenon
of the reversal of the lines.

Kirkwood gaps   Regions in the asteroid zone which have been swept
clear of asteroids by the perturbing effects of Jupiter. They were
named for the American astronomer Daniel Kirkwood.

KL nebula   See Kleinmann-Low nebula.

Klein-Alfvén cosmology   A cosmology in which the observed expansion of the Universe results from the bounce of an originally
collapsing cloud of matter and antimatter. The bounce is caused
by radiation pressure generated by annihilations when the cloud
reaches high density (10^{-2} cm^{-3}).

Klein-Nishina formula   An expression for the total or differential
cross section for the Compton scattering of a photon by a free
electron according to Dirac's electron theory.

Kleinmann-Low nebula (KL nebula)   A cool (< 600 K) extended infrared source in the Orion Nebula, about 1′ NW of the Trapezium and about 12″ south of the BN nebula, discovered in 1967. It

dominates the infrared emission at $\lambda \geq 20$ μ, and a CO cloud is centered on it. It is probably a collapsing cloud of 10^2–10^3 $M \odot$ in which protostars are embedded.

klystron   A type of electron tube used in radar and high-frequency radio work.

knock-on spectrum   A spectrum of particles that are being built up— the inverse of a spallation spectrum.

Knudsen number   The ratio of the mean free path length of the molecules in a fluid to a characteristic length of the structure in the fluid stream.

Kolmogorov-Smirnov test   A nonparametric test used in statistics. The Kolmogorov statistic is simply the magnitude of the maximum deviation between the integral distribution function of a sample and the theoretical distribution one wishes to test.

Kramers's opacity   The opacity of stellar material derived by Kramers, who in 1923 carried out theoretical calculations of stellar opacity as a function of chemical composition.

KREEP (from K for potassium, REE for rare-earth elements, P for phosphorus)   Lunar basaltic material rich in radioactive elements.

Krüger 60 AB   A faint, twelfth-magnitude dM binary ($P = 44.5$ years) in the solar neighborhood (3.93 pc distant). It may be a subluminous star.

Kruskal diagram   A spacetime diagram exhibiting the properties of the Schwarzschild metric by eliminating the formal singularity that appears at the Schwarzschild radius in the form in which the metric is usually written. It exhibits the character of the horizon that appears at the Schwarzschild radius and illustrates the true nature of the singularity at the center. The Kruskal diagram fully describes spacetime in the vicinity of a black hole all the way down to the singularity.

Krzemiński's star   See Cen X-3.

Kuiper bands   Several linelike features near 7500 Å in the spectra of Uranus and Neptune, now identified as methane bands.

kurtosis   The peakedness or flatness in the graphical representation of a statistical distribution.

L

L component   The part of the solar corona whose spectrum consists of emission lines.

L galaxy   In Morgan's classification, an elongated galaxy of low surface brightness.

L-magnitude   The magnitude derived from observations at an infrared wavelength of 3.5 microns.

l-number   The orbital quantum number, which determines the magnitude of an electron's angular momentum.

BL Lacertae   A highly variable object (the most rapid radio variable known, also an optically violent variable—m_v = 12 to 15 mag—and an infrared source). Probably an exceedingly compact nonthermal object, and undoubtedly extragalactic. Its optical spectrum is characterized by an absence of lines, so its redshift cannot be measured. (In 1974 Oke and Gunn infer z = 0.07 from an Hβ absorption line in the surrounding halo and conclude that it lies at the center of a bright (M_V = -23) elliptical. If true, this would make BL Lac the nearest known quasar.)

BL Lacertae object   A member of a class of astronomical objects with the following characteristics: (1) rapid variations in intensity at radio, infrared, and optical wavelengths; (2) energy distributions such that most of the energy is emitted at infrared wavelengths; (3) absence of discrete features in low-dispersion spectra; and (4) strong and rapidly varying polarization at visual and radio wavelengths.

Lagoon Nebula (M8, NGC 6523)   An emission nebula in Sagittarius $\sim$ 2 kpc distant.

Lagrangian points   Five points in the orbital plane of two massive particles in circular orbits around a common center of gravity, where a third particle of negligible mass can remain in equilibrium. Three of the points are on the line passing through the centers of mass of the two bodies—L_2 beyond the most massive body, L_1 (the point through which mass transfer occurs) between the two bodies, and L_3 beyond the less massive body. All three of these points are in unstable equilibrium. The other two (L_4 and L_5) are stable, and are located at the two points in the orbit of the less massive component which are equidistant from the two main components. (See Trojans)

Lallemand camera   A type of image tube (q.v.).

Lamb shift   The difference in energy levels of the H atom between $2S_{1/2}$ and $2P_{1/2}$.

lambda doublet   Two lines in the microwave region of the spectrum of the OH molecule caused by splitting of electronic levels.

Lambert's law (also called cosine law)   The intensity of the light emanating in a given direction from a perfectly diffusing surface is proportional to the cosine of the angle of emission measured between the normal to the surface and the emitted ray.

Landau damping   Damping caused by electrons that are moving at the phase velocity of the wave. It is analogous to a surfer who will be carried along by a wave if he is already moving at the velocity of the wave when it hits him.

Landé factor   The constant of proportionality relating the separations of lines of successive pairs of adjacent components of the levels of a spectral multiplet to the larger of the two J-values for the respective pairs. The interval between two successive components J and $J + 1$ is proportional to $J + 1$.

Lane-Emden equation   A second-order nonlinear differential equation describing the structure of polytropes.

Large Magellanic Cloud (LMC)   See Magellanic Clouds.

Larmor frequency   The frequency of precession of a charged particle orbiting in a uniform magnetic field. It is equal to $eH/4\pi m_e$, where e is the electron charge, m_e is the electron mass, and H is the magnetic field strength.

Larmor radius   The radius of the circular orbit that a charged particle describes transverse to a magnetic field.

laser   A maser (q.v.) which emits radiation at optical wavelengths.

late-type stars   Stars of spectral classes K, M, S, and C.

least squares, principle of   A principle which states that the best estimate of an experimental quantity, deducible from a number of observations, is that for which the sum of the weighted squares of the residuals is least.

Lemaître universe   A big-bang cosmology proposed by the Belgian Abbé Lemaître in 1929 in which the Universe is assumed to have exploded from a primeval "atom." In the Lemaître universe the rate of expansion steadily decreases.

Lennard-Jones potential   An approximation of the interaction between two atoms or molecules.

Lense-Thirring effect   The precession of the plane of the geodesic orbit of a test particle around a rotating mass in general relativity. It arises from the coupling of the rotation of the central mass with the orbital angular momentum of the test particle. This precession is described as resulting from the dragging of inertial frames.

Lenz's law   The current induced by an electromotive force will appear in such a direction that it opposes the charge that produced it.

Leo systems   Two dwarf elliptical galaxies (about 220–250 kpc distant) belonging to the Local Group. Leo I (dE3), $M_V \approx -11$, diameter 1.8 kpc; Leo II, $M_V \approx -9.5$, diameter 1.3 kpc.

17 Leporis   A close binary system (A0 V, M1 III) with a shell-like spectrum indicating that mass transfer may be occurring from the late-type companion onto the A0 primary.

lepton   Any fermion (q.v.) that does not participate in the strong interactions. Leptons include the electron family and the muon family (q.v.).

lepton era   The era following the hadron era (q.v.), when the temperature had dropped to about 10^{12} K and when the Universe consisted mainly of leptons and photons. It started about 10^{-4} s after the big bang and lasted until about 10 s after the big bang; it was followed by the radiation era (q.v.).

libration   Any of several oscillations in the apparent aspect of the Moon as seen from Earth, which, when combined, enable Earth-based observers over a period of time to see altogether about 59 percent of the Moon's surface. Physical librations are angular motions about the center of mass due to gravitational torques on the Moon. Optical librations are the apparent rotations of the Moon, caused by our observing it from somewhat different directions at different times.

libration orbits   See Lagrangian points.

light cone   The set of all directions in which a light signal can travel toward an event (past light cone) or from an event (future light cone).

light curve   A plot of magnitude or intensity versus time for a variable star.

light cylinder   The cylinder whose radius is that at which the rotational velocity of a neutron star would equal the speed of light. $R_L = cT/2\pi$, where R_L is the radius of the light cylinder and T is the period.

light elements   In astrophysics, usually Li, Be, and B.

light pressure   See radiation pressure.

light-year (lt-yr)   The distance light travels in a vacuum in 1 year. 1 lt-yr = 9.4605×10^{12} km = 0.307 pc (c = 299,792.46 km s^{-1} = 186,274 miles s^{-1}). (1 lt-min $\approx$ 0.13 AU.)

limb   Apparent edge of the disk of a solar-system body as projected on the sky (cf. terminator).

limb brightening   Increase in the intensity of radio or X-ray brightness of the Sun or other star from its center to its limb.

limb darkening   Decrease in, and reddening of, the optical brightness of the Sun or other star from its center to its limb.

Lindblad resonance   A resonance hypothesized by Lindblad in the 1920s in his attempt to explain the existence of spiral arms (see density wave theory). It is a resonance which occurs when the frequency at which a star encounters the galactic spiral wave is a multiple of its epicyclic frequency. The inner Lindblad resonance occurs whenever the ratio of the frequency of the radial oscillation to that of the rotational motion around the center of the galaxy (in a frame of reference rotating together with the spiral pattern) is 2:1.

line blanketing   The combined effects of spectral lines upon the emergent energy distribution from and the temperature distribution in a stellar or planetary atmosphere.

line broadening   Increase in the range of wavelengths in which some characteristic emission or absorption occurs, due to a number of causes (e.g., Doppler broadening, the effects of perturbers, etc.).

line profile   A plot of intensity versus wavelength across a spectral line.

line wings   Broad "wings" that appear on either side of a spectral line when the number of atoms producing the line is very great.

Liouville's theorem (the equation of continuity)   For a general Hamiltonian system the volume of an element of phase space is invariant with respect to the equations of motion. As applied to astronomy, the difference between the number of the stars moving into a volume of six-dimensional phase space at a given time and those moving out of it at the same time must be equal to the increase in the distribution function for those stars.

lithium stars   Peculiar giant stars (spectral types G–M) whose spectra show high abundances of lithium. They are primarily S stars and carbon stars, although Li is also found in T Tauri stars, and is sometimes observed in normal late-type giants. Interstellar Li/H $\approx 2 \times 10^{-10}$. (Solar system Li/H $\approx 10^{-9}$.) Lithium is destroyed rapidly (in about 7500 years) at typical nuclear burning temperatures. Spallation is the only production mechanism known for ^{6}Li, but ^{7}Li can be transported from the core in the form of ^{7}Be and converted in the envelope to ^{7}Li by electron capture. Observed ^{7}Li/^{6}Li > 10.

"little bangs"   Hypothetical explosions of supermassive stars shortly after the big bang, with the release of processed elements into the interstellar medium, postulated by Wagoner to account for the anomalously high abundance of helium in the universe, and for the fact that even the oldest stars have some metals.

lobes   In radio astronomy, regions of sensitivity in an antenna pattern. Lobes are analogous to fringes in optical astronomy.

local arm   See Orion arm.

Local Group   The cluster of galaxies to which our Galaxy belongs. It is a poor, irregular cluster with some 20 certain members: three spirals (the Galaxy, M31, and M33); four irregulars (LMC, SMC, IC 1613, and NGC 6822); and about 13 intermediate or dwarf ellipticals (NGC 147; NGC 185; NGC 205; M32; the Sculptor, Fornax, Leo I and II, UMi, and Draco systems; and three companions to M31 discovered by van den Bergh in 1972). It may also include several other dwarf galaxies as well as the giant elliptical Maffei 1. The total mass of the Local Group is less than 1.5 times the combined masses of the Galaxy and M31.

local hypothesis (of quasars)   The hypothesis that quasars are not at the distances inferred from their redshifts.

local standard of rest (LSR)   A frame of reference in which the mean motion of stars in the immediate neighborhood is zero. In such a reference system, the motions of stars in the solar neighborhood (a volume of space about 100 pc in diameter) average out to zero (cf. solar motion). It is a coordinate system in which the origin is a point in the galactic plane moving in a circular orbit around the galactic center with a period equal to that of the Sun, and in which the three velocity components are Π, in the direction from the galactic center to the origin; Θ, in the direction of galactic rotation; and Z, perpendicular to the galactic plane.

Local Supercluster   The supercluster of galaxies to which the Local Group may belong (see Virgo Supercluster). De Vaucouleurs lists 54 groups of galaxies in the LSC ($R \approx 17$ Mpc).

local thermodynamic equilibrium (LTE)   The assumption that all distribution functions characterizing the material and its interaction with the radiation field at a point in the star are given by thermodynamic equilibrium relations at local values of the temperature and density.

Lommel-Seeliger surface   A surface with large-scale roughness where shadowing effects are important.

longitude of the perihelion (ω)   For a solar-system body, the longitude of the ascending node plus the angle along the orbit from the node to the perihelion point.

longitudinal wave   A wave vibrating along the direction of propagation—e.g., a sound wave.

long-period variables (sometimes called red variables)   Pulsating disk-population red giants or supergiants with periods of 100 to 1000 days (Population I typically have periods greater than 200 days;

Population II, periods less than 200 days). Typical is Mira (o Ceti), which has a period of 331 days. Long-period variables may vary by as much as 9 magnitudes in the visible, but in the integrated spectrum (most of their radiation is in the infrared) they vary by only 2 or 3 magnitudes. They are usually of spectral type M, R, or N.

look-back time   The time in the past at which the light we now receive from a distant object was emitted. Galaxies of a certain type (redshift and luminosity) can be seen only at a certain distance.

Loop Nebula   See 30 Doradus.

Lorentz force equation   The equation relating the force on a charged particle to its motion in an electromagnetic field.

Lorentz-invariant   Invariant with respect to Lorentz transformations.

Lorentz transformation   A transformation which enables one to relate the physical parameters describing an object when viewed in one frame of reference to those which are appropriate to an observer moving with a uniform velocity in that frame.

Loschmidt number   Number of molecules of an ideal gas per unit volume (2.687×10^{19} molecules per cm^3).

lossy   Subject to absorption of light. If a material is not lossy, it means that light would be scattered or reflected off.

Love number   A measure of the rotational deformation of a rotating body in hydrostatic equilibrium.

Lowell's band   A dark border sometimes found on the Martian polar cap.

LS coupling (also called Russell-Saunders coupling)   A condition in an atom in which the spins couple to spins and orbital angular momenta couple to orbital angular momenta to form total spin and total orbital angular-momentum vectors which then couple to form the total angular momentum of the atom. In this case spin-spin and orbit-orbit torques exceed spin-orbit torques; the opposite extreme results in j-j coupling.

LSR   See local standard of rest.

LTE   See local thermodynamic equilibrium.

luminosity   Total radiant energy output per second (absolute brightness, usually expressed in ergs s^{-1} or in magnitudes). $L = 4\pi R^2 \sigma T^4$ (see Stefan's law).

luminosity class   A classification of stellar spectra according to luminosity for a given spectral type. The luminosity class is an indication of a star's surface gravity and tells whether it is a dwarf, a giant, or supergiant. Luminosity classes (MKK system): 0, very brightest; Ia, supergiants of high luminosity; Ib, supergiants of low luminosity; II, luminous giants; III, normal giants; IV, subgiants; V, main-sequence stars.

luminosity function   Number distribution of stars or galaxies with respect to their absolute magnitudes. The luminosity function shows the number of stars of a given intrinsic luminosity (or the number of galaxies per integrated magnitude band) in a given volume of space. The stellar density in the solar neighborhood is about 0.16 $M_\odot$ per cubic parsec.

luminous mass   The mass contributed by luminous matter in galaxies (see missing mass). Luminous mass density, 5×10^{-32} g cm^{-3} for $H_0 = 50$ km s^{-1} Mpc^{-1}.

lunation   The period of time between two successive new Moons (cf. synodic month).

lunisolar precession   That component of general precession (q.v.) that is caused by the gravitational coupling between the Moon and the Earth and between the Sun and the Earth. Lunisolar precession causes the equinox to move *westward* along the ecliptic about 50″ per year (cf. planetary precession).

Lupus Loop   A radio source, a large broken shell 4°.5 in diameter, identified as a prehistoric supernova remnant.

Luyten 726-8   A binary (M5.5e V, M6e V) (component B is UV Cet) with a very small mass (total mass of system [1974] 0.3 $M_\odot$). It is about 2.7 pc distant.

Lyman series   The spectral series of the hydrogen atom associated with the first energy level or ground state. The series lies in the ultraviolet (Lα at 1215.67 Å; Lβ at 1026 Å). The series limit is at 912 Å. (He II Lα, 303.78 Å; He II Lyman limit, 227 Å.)

Lyot division (in Saturn's rings)   The gap between rings B and C.

α **Lyrae**   See Vega.

β **Lyrae stars**   A class of eclipsing binaries whose secondary minima are intermediate between those of β Persei and those of W UMa. The prototype β Lyr (B8.5 II, F V) is a complex eclipsing system and is presently in a state of rapid mass transfer. The spectrum of one companion is invisible; it may be a black hole. Beta Lyrae is also a weak radio source.

RR Lyrae stars (also called cluster variables)   A large class of pulsating (amplitude variation about 1 mag) blue giants of anomalous spectral type (A2–F6) with periods of less than 1 day. Their average absolute magnitude is about +0.8, which makes them almost 50 times more luminous than the Sun. They are Population II objects often (but not always) present in globular clusters. RR Lyrae stars are valid distance indicators out to more than 200 kpc.

M

M-magnitude   The magnitude derived from observations at an infrared wavelength of 5 microns.

M star   Stars of spectral type M are cool red stars with surface temperatures of less than 3600 K whose spectra are dominated by molecular bands, especially those of TiO. Examples are Betelgeuse and Antares. M dwarfs are the most numerous type in our Galaxy.

MS stars   M-type stars with ZrO bands.

Mach number (in shock flow)   Ratio of the speed of a shock wave to the sound speed in the same medium.

Mach's principle   A pre-relativity statement to the effect that the local inertial frame is determined by some average of the motion of all the matter in the Universe. In essence, Mach's principle says that space, which is the arena in which matter interacts, is itself an aspect of that matter.

Maclaurin series   A power series expansion of f(x) of the form $f(x) = f(0) + f'(0)x + [f''(0)/2!]x^2 + \ldots + [f^{(n)}(0)/n!]x^n + \ldots$ (= Taylor series at $x = 0$).

Maclaurin spheroid   A form which a homogeneous self-gravitating mass can take when in a state of uniform rotation. Its eccentricity varies from zero (when it is not rotating) to 1 in the limit of infinite angular momentum.

Maffei 1 and 2   Two galaxies discovered on infrared plates in 1968 and identified in 1970 as members (probably temporary) of the Local Group. The two galaxies lie in the zone of avoidance (q. v.). Maffei 2 has since been classified as a medium-sized, average luminosity Sbc II spiral at a distance of 5 ± 2 Mpc—too far away to belong to the Local Group—but Maffei 1 (a giant elliptical) is probably only ~1 Mpc distant, marginally within the Local Group.

Magellanic Clouds   Two small irregular (or possibly barred spiral) galaxies (satellites of the Milky Way Galaxy) about 50–60 kpc (LMC, in Dorado) and 60–70 kpc (SMC, in Toucan) distant, visible to the naked eye from the southern hemisphere. Both clouds contain mainly Population I stars. The LMC contains numerous OB stars and at least 10 stars that are an order of magnitude brighter ($M_V = -9$) than any supergiants known in our Galaxy. It also contains several times our Galaxy's concentration of interstellar matter.

Magellanic Stream   A name given to the long H I filament that extends from the region between the Magellanic Clouds down to the south galactic pole and which appears to make a 180° arc of a great circle across the sky.

Magellanic System   A name given to the common envelope of neutral hydrogen in which the LMC and SMC are embedded.

magnetic bremsstrahlung   See synchrotron radiation.

magnetic-dipole radiation   Radiation emitted by a rotating magnet.

magnetic moment   The ratio of the torque experienced by a magnet aligned perpendicular to a magnetic field to the strength of the magnetic field.

magnetic pressure   The pressure exerted by a magnetic field on the material that contains the field. In Gaussian units it is given by $p_m = B^2/8\pi$, where B is the magnetic field strength.

magnetic stars   Stars (usually of spectral type A) with strong integrated magnetic fields ranging up to 30,000 gauss.

MHD   (magnetohydrodynamics; sometimes called hydromagnetics)   The study of the collective motions of charged particles in a magnetic field.

magneton   See Bohr magneton.

magnetopause   The region in Earth's ionosphere where the magnetosphere meets the solar wind. Essentially, it is the place where Earth's magnetic field stops; the region above the magnetopause is no longer part of Earth's atmosphere, but is part of interplanetary space.

magnetosphere   The region of space surrounding a rotating, magnetized sphere. Specifically, the outer region of Earth's ionosphere, starting at about 1000 km above Earth's surface, and extending to about 60,000 km (or considerably farther, on the side away from the Sun).

magnitude   An arbitrary number, measured on a logarithmic scale, used to indicate the brightness of an object. Two stars differing by 5 mag differ in luminosity by 100. 1 magnitude is the fifth root of 100, or about 2.512. The brighter the star, the lower the numerical value of the magnitude (see also Pogson's ratio).

Maia sequence   A hypothetical sequence of blue variable stars named for the B-type star Maia (20 Tau) in the Pleiades. Maia has the lowest rotational velocity of any B star in the Pleiades.

main beam   The lobe of maximum sensitivity in a radio telescope.

main lines (of an OH source) The transitions that emit radiation at 1665 and 1667 MHz.

main sequence The principal sequence of stars on the H-R diagram, containing more than 90% of the stars we observe, that runs diagonally from the upper left (high temperature, high luminosity) to the lower right (low temperature, low luminosity). A star appears on the main sequence after it has started to burn hydrogen in its core, and is estimated to stay on the main sequence until it has used up about 12% of its hydrogen (for a 1 $M\odot$ star, about 10^{10} yr). The observed upper limit of the main sequence is 60 $M\odot$; stars above this limit are unstable to nuclear driven pulsations in the early stages of stellar evolution. The calculated lower limit of the main sequence is 0.085 $M\odot$; stars below this limit cannot achieve nuclear chain reactions. H to He fusion represents more than 80% of the maximum possible nuclear energy available to a star.

Maksutov telescope A reflector whose primary mirror is spheroidal instead of parabolic. The light initially passes through a large concave lens to remove the spherical aberration.

Malmquist correction A correction introduced into star counts distributed by apparent magnitude.

manganese stars Stars with an anomalously high Mn-Fe ratio, which show deviations from the odd-even effect for phosphorus, gallium, and yttrium.

manifold A mathematical concept used to describe the geometry of spacetime.

mantissa The decimal part of a common logarithm.

marching subpulse The weaker component of a pulsar pulse when its period is more than half that of the main pulse, so that the subpulse occurs at progressively later intervals between successive main pulses.

mare (pl. maria) An area on the Moon that appears darker and smoother than its surroundings. Lunar maria are scattered basaltic flows.

Markarian galaxy A galaxy in Markarian's list of galaxies with abnormally strong ultraviolet continua. They have broad emission lines arising in a bright, semistellar nucleus. Markarian 231 is the most luminous galaxy known if it is at its Hubble distance.

Mars Fourth planet from Sun. Mass 6.45×10^{26} g (0.11 Earth's); radius (1974) 3394 km. Oblateness 0.0092. Mean density 4.0 g cm^{-3}. Mean distance from Sun 1.5237 AU; $e = 0.0934$; $i = 1°.85$. Sidereal period 687 days; synodic period 779.9 days.

Mean orbital velocity 24.2 km s^{-1}. Surface temperature 248 K. Rotational period $24^h37^m22^s6$. Obliquity 23°59′. Surface gravity 0.38 that of Earth; escape velocity 5.1 km s^{-1}. Albedo 0.16. Atmosphere more than 90% CO_2, traces of O_2, CO, H_2O. Atmospheric pressure from *Mariner 7*, 3.5 millibars. The core is probably liquid Ni-Fe. Two tiny satellites (Phobos and Deimos), both of which are locked in synchronous rotation with Mars.

mascons (mass concentrations)   Gravity anomalies found on the Moon. As of 1971, 12 mascons were known.

maser (*m*icrowave amplification by stimulated emission of radiation)   A device that utilizes the natural oscillations of an atom or molecule to amplify electromagnetic radiation. Molecules are pumped into a metastable upper state by incident radiation of broad frequency via upper states that connect to the metastable state. They are then stimulated downward by radiation of a certain frequency connecting the metastable and ground states. When a bound electron in the metastable state is hit by a photon of the right frequency, the electron can return to a lower state by emitting a photon of exactly the same frequency as the incident photon; and it will emit it in exactly the same direction in which the incident photon is scattered. This means that the photons move off precisely in phase. If each hits another electron in the same state, there will be four photons in phase, etc.

mass absorption coefficient   A measure of the fraction of radiation absorbed at a certain wavelength per unit mass.

mass defect   The deficiency in the mass of an atomic nucleus with respect to the sum of the individual masses of its constituent particles. It represents the amount of mass converted into the binding energy of the nucleus; and when the nucleus is formed, this binding energy is released. The Sun is losing more than 4 million tons of mass every second by the conversion of hydrogen into helium.

mass discrepancy   In the study of clusters of galaxies, the difference between the mass of a cluster obtained by using the virial theorem and the mass (usually much smaller) inferred from the total luminosities of the member galaxies. Typically $M_{VT}/M_L > 10$.

mass fraction   The fractional amount (by mass) of a given element or nuclide in a given composition.

mass function   A numerical relation between the masses of the two components of a spectroscopic binary when the spectral lines of only one component can be seen: $f(M_p, M_s) = (M_s^3 \sin^3 i)/(M_p$

+ $M_s)^2$, where M_p = mass of primary, M_s = mass of secondary, and i = inclination of the orbit.

mass-luminosity relation   Plot, on a logarithmic scale, of bolometric absolute magnitude versus mass for main-sequence stars of given chemical composition, derived by Eddington in 1924: $(L/L\odot) = (M/M\odot)^\alpha$, where in general $\alpha \approx 3.5$.

mass-luminosity-radius relation   All nondegenerate stars with the same mass and the same chemical composition will have the same radius and the same luminosity. (See Vogt-Russell theorem.)

mass-radius relation (Chandrasekhar)   For any given mass less than the Chandrasekhar limit (q.v.), there is a unique radius for a totally degenerate star.

matrix   A rectangular array of numers or algebraic quantities representing a system of entities related in a systematic manner. Matrices do not obey the commutative laws of multiplication. An $m \times n$ matrix has m rows and n columns. Matrices may often by abbreviated to $A = [a_{ij}]$. By convention, the first subscript (i) gives the number of the row; the second (j) gives the number of the column.

matter era   The era following the radiation era (q.v.). The matter era started when the temperature of the primeval fireball had dropped to 3000 K, at which time the recombination of hydrogen became possible.

maxwell   The cgs unit of magnetic flux through 1 cm^2 normal to a field of 1 gauss. 1 Mx corresponds to 10^{-8} Wb.

Maxwell distribution   An expression for the statistical distribution of velocities among the molecules of a gas at a given temperature.

Maxwell-Boltzmann distribution   The distribution function that any species of particle will have if it is in thermodynamic equilibrium. This distribution function describes both the equilibrium in velocity space or kinetic energy, and the equilibrium in potential energy.

Maxwell's equations   Equations governing the varying electric and magnetic fields in a medium.

Me stars   Stars of spectral type M with emission lines in their spectra.

mean free path   Mean length of the path of a particle between collisions.

mean life   The mean time before decay of a large number of similar particles. It is equal to 1.44 times the half-life.

mean molecular weight (μ)   Total atomic (or molecular) weight divided by the total number of particles. For instance, the mean molecular weight of a plasma of pure ionized ^{4}He would be 4 (the atomic mass number) divided by 3, the total number of particles (1 nucleus plus 2 electrons). So μ would equal $\frac{4}{3}$.

mean profile (of a pulsar) (also called integrated profile or pulse window)   The relatively stable curve obtained by synchronously averaging together many pulses.

mean solar day   The mean length of time ($24^h00^m00^s$) between two successive culminations of the Sun (i.e., the mean period from apparent noon to apparent noon).

mean solar second   1/86,400 of a mean solar day (cf. ephemeris second).

mean Sun   A fictitious body that moves eastward in a circular orbit along the celestial equator, making a complete circuit with respect to the vernal equinox in a tropical year.

mega-   A prefix meaning 10^6.

Meinel bands   Molecular bands of the N_2^+ radical near 8000 Å.

Mercury   Innermost planet of the solar system. Mass (from *Mariner 10*) 3.15×10^{26} g (0.054 Earth's); equatorial radius 2446 km; mean density 5.44 g cm^{-3}. V_{esc} 4.2 km s^{-1}; surface gravity 360 cm s^{-2}. Mean distance from Sun 0.387 AU; $e = 0.206$, $i = 7°.0$. Orbital period 88 days; V_{orb} 47.9 km s^{-1}. Rotation period 58.646 days, exactly $\frac{2}{3}$ of its orbital period. Oblateness <0.001. Synodic period 116 days. Albedo 0.06. Maximum elongation 28°. Transits of the Sun occur either 7 or 13 years apart—last transit 1973 November 10. Advance of perihelion 476″ per century; relativistic advance 42″.6 per century. Subsolar point temperature (from infrared observations) 408 ± 8 K; from *Mariner 10*, 700 K. Optical spectrum is similar to lunar maria and uplands. Surface is probably covered with lunar-like soil of low-density silicates (the core must be rich in iron to account for the high density and the magnetic field). *Mariner 10* has established the existence of an extremely thin ($< 2 \times 10^{-9}$ millibars) atmosphere of He and Ar, and also the presence of a weak magnetic field (200–300 gammas), apparently inclined $<10°$ to the pole and offset 0.47 R_M.

meridian   The great circle passing through the zenith of the observer and the north and south points on his horizon.

meridional flow   Flow between the poles, or between the equator and the poles. A positive value indicates flow away from the equator; a negative value, flow toward the equator.

meson   A nuclear particle (see boson) with a mass intermediate between that of a proton and an electron, which is believed to be responsible for the strong nuclear force. Mesons have a spin of 0. In contrast to the case of baryons or leptons, meson number is not conserved; like photons, mesons can be created or destroyed in arbitrary numbers. Their charge can be positive, negative, or zero.

mesosphere   The part of Earth's atmosphere immediately above the stratosphere, where the temperature drops from about 270 K to 180 K.

Metagalaxy   A synonym for the Universe.

metal   As used in stellar spectroscopy, any element heavier than helium.

metal-enhanced star formation   A hypothesis according to which stars form preferentially from regions of above-average Z in a chemically inhomogeneous interstellar medium.

metallic hydrogen   A hypothetical form of hydrogen in which the molecules have been forced by extremely high pressures to assume the lattice structure typical of metals. It is estimated that as much as 40% of Jupiter's mass (but not more than 3% of Saturn's) may be in the form of metallic hydrogen.

metal-rich stars   Stars having metal-to-hydrogen ratios greater than those of the Hyades.

metastable state   An excited state in an atom, of relatively long duration, from which the atom, under laboratory conditions, cannot pass directly to the ground state by emitting radiation. Under laboratory conditions, the mean time between collisions is so much shorter than the lifetime of the metastable state that an atom is virtually certain to collide with another atom and lose its energy by collision before it has had a chance to radiate. In the extremely rarefied interstellar medium, however, such highly improbable transitions do occur, and the spectral lines emitted by such transitions are called forbidden lines.

meteor   A "shooting star"—the streak of light in the sky produced by the transit of a meteoroid through the Earth's atmosphere; also the glowing meteoroid itself. The term "fireball" is sometimes used for a meteor approaching the brightness of Venus; the term "bolide" for one approaching the brightness of the full Moon.

meteor shower   A profusion of meteors that fall within a period of a few hours and that appear to radiate from a common point in the sky. Shower meteors are usually low-density material, have high eccentricities, and are known to be associated with comets (e.g., the orbit of the Leonids is identical with that of comet Tempel-Tuttle 1866 I).

meteorite   A solid-body portion of a meteor that has reached Earth's surface. Meteorites are divided into three main classes: aerolites (stony meteorites), siderites (iron meteorites), and siderolites (stony iron meteorites). Most meteorites are high-density objects related to asteroids. They range in size from subplanetary to microscopic. Several hundred tons of meteoritic material are estimated to fall on Earth daily.

meteoroid   A small particle orbiting the Sun in the vicinity of Earth.

meter   The SI unit of length. The meter is the length equal to 1,650,763.73 wavelengths in vacuum of the radiation corresponding to the transition between the levels $2p_{10}$ and $5d_5$ of the Krypton-86 atom (11th CGPM, 1960, Resolution 6).

methanol (CH_3OH)   An organic molecule (also called methyl alcohol) discovered in interstellar space in 1970. More rotational lines have been observed astronomically for it than for any other molecule.

methylamine (CH_3NH_2)   A molecule discovered in interstellar space in 1974, in Sgr B2, at a frequency of 87.77 GHz. Methylamine can react with formic acid to produce glycine, the simplest amino acid.

metric   The metric gives the spacetime interval ds between two neighboring events.

micro-   A prefix meaning 10^{-6}.

micron (μ)   A unit of length equal to 10^{-4} cm, or 10^4 angstroms.

microphotometer   A device for measuring the variations in density in a photographic emulsion.

microwave   An electromagnetic wave (in the radio region just beyond the infrared) with a wavelength of from about 1 mm to 30 cm (about 10^9–10^{11} Hz).

microwave background   See cosmic background radiation.

Mie scattering   Scattering of light (without regard to wavelength) by larger particles, such as those of dust or fog in Earth's atmosphere (cf. Rayleigh scattering).

Mie theory   A theory of the diffraction of light by small spherical particles.

Milky Way   A faint band of light around the sky, composed of vast numbers of stars and nebulae lying near the galactic plane. Also, the name of our Galaxy.

milli-   A prefix meaning 10^{-3}.

Mills cross   An antenna array consisting of two antennas oriented at right angles to each other. It produces a single narrow pencil beam.

Milne-Eddington approximation   A first approximation in the analysis of stellar spectra, in which a line is assumed to be formed in such a way that the ratio of the line absorption coefficient to the continuous absorption coefficient is constant with depth. It is used primarily in analyses of the lines of ionized metals, for which cases it is often an accurate approximation.

Mimas   The second innermost satellite of Saturn, discovered by Herschel in 1789. $P = 0\overset{d}{.}94$; $R \approx 250$ km. Albedo 0.49. It is the perturbations of Mimas and Janus that produce the divisions in Saturn's rings.

minimum resolvable angle   In radians, 1.22λ divided by the aperture of the telescope.

mini-quasar   Sandage's term for the blue nucleus of an N galaxy. According to Sandage, N galaxies can be understood as ordinary giant elliptical galaxies with "mini-quasars" embedded in them.

Minkowski space   A four-dimensional spacetime with a flat (i.e., Euclidean) geometry, which is used in the special theory of relativity.

minor planet   See asteroid.

Mira (o Ceti)   A M6e–M9e III irregular long-period intrinsic variable about 70 pc distant. Mean period 331 days. It can be more than 5 mag brighter at maximum than at minimum. It was named Mira ("wonderful") in 1596 by Fabricius, who made the first recorded observations of its brightness fluctuations. Mira is a double star with a faint B companion which is itself variable.

Miranda   The innermost satellite of Uranus, discovered by Kuiper in 1948. $P = 1^{d}10^{h}$, diameter about 500 km.

"missing mass"   See mass discrepancy.

mixing-length theory   A semiempirical theory used to describe convection phenomena in stars.

MKK system   A classification of stellar spectra according to luminosity, devised by Morgan, Keenan, and Kellman (see luminosity class). The MKK system uses two parameters (spectral type and luminosity class) to describe a system with three variables (temperature, luminosity, and abundance).

mole (mol)   The SI unit of the amount of substance, defined as the amount of substance of a system which contains as many elementary entities as there are atoms in 0.012 kilograms of carbon 12 [14th CGPM 1971, Resolution 3]. 1 mole, which is equal to 1 gram multiplied by the molecular weight, contains 6.02×10^{23} molecules (see Avogadro's number). In general, 1 mole of any gas occupies a volume of 22.4 liters.

molecular hydrogen (H₂)   A molecule of hydrogen, discovered in interstellar space in 1970. H_2 is a very hard molecule to detect. None of its transitions lie in the visible part of the spectrum. Second, being a symmetric homonuclear molecule, it does not have an electric-dipole rotation-vibration spectrum, and detection must be based on the weak quadrupole spectrum. Third, ultraviolet radiation is a very efficient dissociator of H_2, so any H_2 that survived would presumably be located inside very dense interstellar clouds. So far observations have borne out this supposition. Measurements of the region within about 1 kpc of the Sun suggest that H_2 is about twice as abundant as atomic H.

moment of inertia   The product of the mass of a body and the square of its radius of gyration.

Monoceros Loop   A filamentary loop nebula about 1 kpc distant, the remnant of a supernova that occurred about 300,000 years ago.

R Monocerotis   An A–F pec variable star that illuminates the variable cometary nebula NGC 2261. Its temperature is about 810 K, and it is a source of CO emission.

monodromy   The property that all paths of points of a body simply rotating about an axis shall return into themselves.

Monte Carlo method   A trial-and-error technique used on computers to solve complex problems.

Moon   Natural satellite of Earth. Mass 7.35×10^{25} g $= 0.0123$ Earth's; mean radius 1738 km; mean density 3.34 g cm^{-3}; mean distance from Earth 384,404.377 $\pm$ 0.001 km (1.28 lt-sec); V_{esc} 2.38 km s^{-1}; surface gravity 162.2 cm s^{-2} $= 0.165$ Earth's. Sidereal period $27^d7^h43^m11^s$, $e = 0.0549$, inclination of orbital plane to ecliptic $5°8'43''$. Obliquity $6°41'$. Synodic period $29^d12^h44^m2^s9$. $V_{orb} = 1.02$ km s^{-1}. Albedo 0.07. The Moon's center of mass is displaced about 2 km in the direction of Earth. Studies of lunar rocks have shown that melting and separation must have begun at least 4.5×10^9 years ago, so the crust of the Moon was beginning to form a very short time after the solar system itself. Thickness of crust, $\sim$60 km; of mantle, $\sim$1000 km. Temperature of core, $\sim$1500 K. It would have taken only 10^7 years to slow the Moon's rotation into its present lock with its orbital period. The Moon's orbit is always concave toward the Sun.

Mössbauer effect   Recoil-free gamma-ray resonance absorption.

mottle   An alternative word for spicule.

moving cluster   A group of stars dynamically associated so that they have a common motion with respect to the local standard of rest. Examples are the Hyades and the Ursa Major group.

Mulliken bands   Spectral bands of the C_2 radical.

multiplet   A group of spectral lines arising from transitions having a common lower energy level. The group of lines have the same values of L and S but different values of J.

muon (μ)   An elementary particle, formerly called a mu-meson but now classified with the leptons because it seems to be identical with the electron except for its much greater mass (207 times that of an electron). The muon family includes the muons and their neutrinos (and their antiparticles). Muons may have a positive or a negative charge.

Murchison meteorite   A type II carbonaceous chondrite which fell in 1969 near Murchison, Australia, and which was found to contain at least 17 amino acids. Left-handed and right-handed forms were present in roughly equal quantities.

Murray meteorite   A type II carbonaceous chondrite that fell near Murray, Kentucky, in 1950.

N

N-magnitude   The magnitude derived from observations at a wavelength of 10 microns.

N galaxy   A galaxy with a small, bright, blue nucleus superposed on a considerably fainter red background. (In the Yerkes 1974 system, a galaxy with a small nucleus containing a considerable fraction of the luminosity; N^-, less pronounced N galaxies; N^+, extreme examples of N galaxies.) Also, a type of radio galaxy having a brilliant, starlike nucleus containing most of the luminosity of the system. N galaxies are compact galaxies, and as a class are intermediate between Seyfert galaxies and quasars in properties of form, color, spectra, redshift, and optical and radio variability.

N lines   Two green forbidden lines of doubly ionized oxygen [O III]. N1 is at 5007 Å; N2 at 4959 Å.

N star   Stars of spectral type N are very red stars similar to M stars except that bands of C_2, CN, and CH are present instead of those of TiO. N stars are strongly concentrated toward the galactic plane.

naked singularity   A singularity (q.v.) that will be visible and communicable to the outside world.

nano-   A prefix meaning 10^{-9}.

natural line broadening   The broadening of spectral lines resulting from the fact that excited levels have certain mean lives, and these mean lives, by virtue of the uncertainty principle, imply a spread in the energy values.

near zone (of a pulsar)   The zone of a pulsar within the velocity-of-light radius (q.v.).

nebula   See diffuse nebula, gaseous nebula, dark nebula. The term "nebula" was previously applied to all kinds of hazy patches in the sky, many of which are now recognized to be clusters or galaxies.

nebular lines   Lines that originate from forbidden transitions (q.v.).

negative hydrogen ion   See H^- ion.

negatron   A negatively charged electron, as opposed to a positron.

Neptune   Eighth planet from Sun. Mean distance from Sun 30.07 AU, $e = 0.009$. Orbital period 164.8 years, $V_{orb} = 5.43$ km s^{-1}. Synodic period 367.49 days. Albedo 0.62. Maximum apparent brightness $+7.6$ mag. Mass 1.03×10^{29} g; radius 24,500 $\pm$ 500 km; mean density 1.7 g cm^{-2}. Oblateness 0.02; V_{esc} 25 km s^{-1}; surface gravity 1.3 that of Earth. Rotation period 15^{h}49^{m}30^s. Inclination 1°.8; obliquity 28°.8. The primary constituents of its atmosphere are hydrogen and methane. Two satellites, Triton and Nereid.

Nereid   The outer satellite of Neptune (radius 150–250 km). Period about 360 days direct. It has the most eccentric orbit ($e = 0.76$) of any natural satellite. Discovered by Kuiper in 1950.

Nernst theorem (third law of thermodynamics)   All substances have zero entropy at 0 K.

Neumann lines   In iron meteorites, groups of very fine parallel lines that cross each other at various angles. Irons containing Neumann lines can easily be cleaved in three mutually perpendicular directions.

neutral currents   Interactions between neutrinos and nucleons which involve the exchange of a neutral (i.e., uncharged) intermediary.

neutral region (solar)   A region where the magnetic field strength approaches zero. Generally, neutral regions occur between regions of opposite polarity.

neutrino (ν)   A stable particle with no charge, a rest mass of zero, and a spin of ½, that carries away energy in the course of nuclear reactions. Its main characteristic is the weakness of its interactions with all other particles. Since the wavelengths of neutrinos at the energies at which they are normally emitted from unstable nuclei are only a few thousandths of an angstrom (compared with the wavelength of a light photon which is several thousand angstroms), they have negligible probability (10^{-19} that of a light photon) of interacting with matter and escape at the speed of light. Neutrinos arise only in the energy-producing regions of stars and therefore, unlike light photons, provide direct evidence of conditions in stellar cores. There are two types of neutrinos, those associated with electrons (ν_e) and those associated with muons (ν_μ).

neutrino bremsstrahlung   The reaction in which an electron scatters from a nucleus, emitting a neutrino-antineutrino pair.

neutron   A nuclear particle with zero charge and with a mass slightly greater than that of a proton (mass of neutron 1.008665 amu = 1.6749×10^{-24} g). A free neutron decays, after a half-life of about 10.6 minutes, into a proton, an electron, and an antineutrino. The neutron is probably made up of still more fundamental particles having both positive and negative charges. The charges balance to give a net charge of zero, but the motions of the charges are such that their magnetic contributions do not cancel and consequently the neutron is magnetic.

neutron drip   The rapid increase in the abundance of free neutrons that occurs when physical conditions are such that the neutron becomes the stable nucleon with respect to electron capture (as will happen in a degenerate assembly of electrons with sufficiently high Fermi threshold energy).

neutron excess (η)   The excess of neutrons over protons in an atomic nucleus. $\eta = (N_n - N_p)/(N_n + N_p)$.

neutron matter   Degenerate matter in which the electron:proton:neutron ratio is about 1:1:8 (or perhaps 1:1:10 or 1:1:12).

neutron star   A star whose core is composed primarily of neutrons, as is expected to occur when the mean density is in the range 10^{13}–10^{15} g cm^{-3}. Under current theories pulsars are assumed to be rotating magnetic neutron stars. A neutron star would probably be only 10–15 km in diameter with a magnetic field of about 10^{12} gauss, a density of 10^{13}–10^{15} g cm^{-3} (compared with a white dwarf's maximum density of about 10^8 g cm^{-3}), and a central temperature of about 10^9 K, and thus would be both bluer and dimmer than a white dwarf.

newton (N)   The SI derived unit of force, equal to the force necessary to give an acceleration of 1 meter per second^2 to a mass of 1 kg. 1 N = 10^5 dynes.

Newton's laws   (1) A body remains in a state of rest or uniform motion when left to itself. (2) The net force on a body is equal to the product of its mass and its acceleration. (3) When two bodies interact, the force on the first due to the second is equal and opposite to the force on the second due to the first.

Ney-Allen Nebula   An extended infrared source in the Trapezium region of Orion, which shows a strong 10-μ emission feature assumed to result from circumstellar shells of silicate dust.

Nicol prism   A device made from a split crystal of Iceland spar with which plane-polarized light can be detected.

nightglow   See airglow.

night-sky light   The faint, diffuse glow of the night sky. It comes from four main sources (q.v.): airglow, diffuse galactic light, zodiacal light, and the light from these sources scattered by the troposphere (see also gegenschein, cosmic light).

node   The point at which a standing wave pattern intersects the horizontal axis and at which the wave consequently has zero amplitude.

nodes, line of   The intersection between the orbital plane of the Moon or a planet and the plane of the ecliptic.

nodical month (also called draconic month)   The interval of time (27.2122 days) between two successive transits of the Moon through its ascending node.

noise source   An electronic device designed to generate known amounts of radio noise in order to test and calibrate the receivers of radio telescopes.

noncoherent scattering   Absorption of a photon and reemission at a different frequency (as seen by an observer) by scattering atoms. The natural width of the lines, Doppler broadening, and pressure broadening are the main processes that give rise to noncoherent scattering.

nonconservative scattering   Scattering that occurs in the presence of absorption.

nongray atmosphere   A model atmosphere constructed by letting the absorption coefficient vary with frequency.

nonrelativistic zone (of a pulsar)   The region far from the star where $r \gg M$.

nonthermal radiation   Radiation emitted by energetic particles for reasons other than high temperature of the source. The spectrum of nonthermal radiation is different from that predicted by Planck's law for a blackbody.

Nordtvedt effect   A theoretical violation of the principle of equivalence (q.v.) for massive, self-gravitating bodies.

normal modes   All the characteristic frequencies of an oscillating body.

normalization   A mathematical technique for eliminating divergent terms or for making them converge.

north point   The point at which the meridian intersects the horizon below the north celestial pole.

North America Nebula   An emission nebula (NGC 7000) in Cygnus.

north polar sequence   A series of accurately measured magnitudes (down to $m_v = 20$) of stars within 2° of the north celestial pole, which have been used to establish an arbitrary zero point on the magnitude scale. (Nowadays the *UBV* system is more frequently used to establish the zero point.)

North Polar Spur   A radio continuum feature extending from the galactic plane to the vicinity of the north galactic pole. It is probably about 50–200 pc distant, and is believed to be a supernova remnant. It is also an X-ray source.

Northern Cross   A group of bright stars in the constellation Cygnus.

nova   A star that exhibits a sudden surge of energy, temporarily increasing its luminosity by as much as 14 mag. (Since 1925 novae have been given variable star designations.) Novae are old disk-population stars. Unlike supernovae, novae retain their stellar form and most of their substance after the outburst. All known common novae are found in close binary systems with one component a cool red giant and the other a hot, less massive object which is the seat of the instability. Starrfield (1974) finds that a C-O core is required for a nova.

nuclear density   The density of an atomic nucleus (about 10^{14} g cm^{-3}).

"nuclear disk"   A rotating disk of about 10^6 $M\odot$ of neutral hydrogen in the inner 800 pc of our Galaxy.

nuclear matter   Matter in which the numbers of protons and neutrons are roughly equal, as in atomic nuclei. Nuclear matter is probably in a liquid or a solid state.

nuclear time scale   Time required for a star to evolve a significant distance off the main sequence; the time it takes a star to convert all its available hydrogen into helium. For the Sun, it is 10^{10} years. (Cf. Kelvin time scale.)

nucleon   A proton or a neutron.

nuclear statistical equilibrium   Equilibrium with respect to strong and electromagnetic interactions.

nucleosynthesis   Formation of atomic nuclei as a result of nuclear reactions in stellar interiors. From the relative abundances of ^{244}Pu, Clayton (1972) has determined that nucleosynthesis began in our Galaxy 12 ± 2 billion years ago. Nucleosynthesis in the average galaxy began 2×10^9 years after the big bang if $H_0 =$ 55 km s^{-1} Mpc^{-1}.

nucleus (of atom)   The massive, positively charged central part of an atom, composed mainly of protons and neutrons, around which the electrons revolve. The radius of an atomic nucleus is directly proportional to the cube root of its mass. Density at least 10^{14} g cm^{-3}. Radius 10^{-12}–10^{-13} cm.

nucleus (of a comet)   The stellar-appearing frozen core, containing almost the entire cometary mass, in the head of a comet.

nuclide   A species of atomic nucleus, analogous to the word "isotope" for a species of atom. The word is also used to distinguish between atomic nuclei that are in different energy states.

null geodesic   The path of a light ray in curved spacetime. It is characterized by the fact that its tangent **U** at any point is of zero length: $\mathbf{U}^\mu \mathbf{U}_\mu = 0$.

number density (n)   Number of particles per cm^3 (cf. column density).

nutation   A small, irregular oscillation in the precessional motion of Earth's rotational axis, caused primarily by lunar perturbations. It has a principal period of 18.6 years, and moves the equinox as much as 17″ ahead of or behind its mean position.

O

O **magnitude**   The magnitude derived from observations at 11 microns.

O star   Stars of spectral type O are very hot blue stars with surface temperatures of about 35,000 K, whose spectra are dominated by the lines of singly ionized helium (see Pickering series). (Most other lines are from at least doubly ionized elements, though H and He I lines are also present.) O stars are useful because they are found in dust clouds and virtually define the spiral arms. Most O stars are very fast rotators. O stars have lifetimes of only 3 to 6 million years.

OB associations   Associations (q.v.) of stars of spectral type O–B2. About 20 are known.

Oef stars   Early O stars that show double emission lines in He II λ4686.

Of stars   Peculiar O stars in which emission features at λλ4634–4641 from N III and 4686 from He II are present. They have a well-developed absorption spectrum, which implies that the excitation mechanism of the emission lines is *selective*, unlike that of Wolf-Rayet stars. The spectra of Of stars are usually variable, and the intensities of their emission lines vary in an irregular manner. Of stars belong to extreme Population I. All O stars earlier than O5 are Of.

OH (hydroxyl radical)   An interstellar molecule first detected in 1963 at a wavelength of 18 cm. The four transitions that occur near 18 cm are caused by the splitting of the ground level. Galactic OH sources have been divided into three classes according to whether the OH emission is strongest in the main lines, particularly at 1665 MHz (class 1), whether the emission and absorption are highly anomalous only in the satellite lines (class 2) (class 2a, 1720-line emitters; class 2b, 1612-line emitters), or whether there is only absorption in all four lines (class 3).

Oberon   Outermost satellite of Uranus, discovered by Herschel in 1787. $P = 13.46$ days (rotational and orbital); $R \geq 500$ km.

objective   The primary mirror of a reflecting telescope (or the primary lens of a refractor).

objective grating   A coarse grating placed in front of the telescope objective.

objective prism   A small-angle prism placed in front of a telescope objective to transform each star image in a field of stars into an image of its spectrum.

oblateness   Ratio of the difference between the equatorial and polar radii to the equatorial radius. Oblateness usually is an indication of how fast the body is rotating.

oblique rotator   A stellar model in which the rotational and magnetic axes are not coincident. Magnetic stars are generally assumed to be oblique rotators of this kind.

obliquity   The angle between a planet's axis of rotation and the pole of its orbit (cf. inclination). It is obliquity that is responsible for the seasons of a planet. Earth's obliquity decreases 0.47 per annum.

occultation   The cutoff of the light from a celestial body caused by its passage behind another object (cf. eclipse). (Strictly speaking, a solar "eclipse" is a solar occultation.)

Ockham's razor   *Entia non sunt multiplicanda* ("Entities are not to be multiplied"). A doctrine formulated by William of Ockham in the fourteenth century. As used by physicists, it means that any hypothesis should be shorn of all unnecessary assumptions; if two hypotheses fit the observations equally well, the one that makes the fewest assumptions should be chosen.

octave   The span over which the frequency doubles; e.g., Middle C is 262 cycles per second; the C one octave above it is 524 cycles per second. The observed electromagnetic spectrum covers a range of 17 decades (about 56 octaves)—from about 10^6 to about 10^{23} cycles per second.

oersted (Oe)   Unit of magnetic field strength. 1 Oe corresponds to $1000/4\pi$ amperes per meter.

Olbers' paradox   A paradox formulated by the German astronomer Heinrich Olbers in 1826: Why is the sky dark at night? The amount of light we receive from a star decreases as the square of the distance from us. On the other hand, if we assume a uniform distribution of stars in space, the number of stars increases as the square of their distance from us, so the two factors should cancel out. In theory, then, the night sky should be a blazing mass of light, and obviously it is not. This self-contradictory statement is Olbers' paradox. In seeking to resolve it, astronomers noted that, besides the assumption of uniformity or homogeneity, Olbers made four other assumptions: (1) space is Euclidean; (2) the laws of physics that apply on Earth apply to the Universe as a whole; (3) the Universe is static (i.e., neither expanding nor contracting); (4) the Universe is spatially and temporally infinite. It is now known that all four of these assumptions are either incorrect or inaccurate.

Omega Nebula (also called Swan Nebula) (M17, NGC 6618)   A bright H II region in Sagittarius. It is a double radio source.

Oort clouds   H I regions extending to more than 100,000 AU from the Sun, barely gravitationally bound, postulated as the birthplace of comets.

Oort's constants (*A* and *B*)   "Constants" that characterize the rotation of our Galaxy with respect to the Sun. A = 0.015 km s^{-1} pc^{-1}; $B = -0.010$ km s^{-1} pc^{-1}.

Oosterhoff groups   Two groups of globular clusters which differ in the period of transition between Bailey type *ab* and type *c* variables, the ratio of type *c* to type *ab* stars, in the metallicity of RR Lyrae stars, and in the mean period of the *ab* variables. On the whole group I clusters have *slightly* weak metal lines whereas group II clusters have *very* weak metal lines.

opacity   A measure of the ability of a gas to absorb radiation. Since the opacity at a given temperature depends on the number of particles per unit volume and since heavier elements contain more electrons than lighter elements, the opacity of a star will increase with increasing proportions of heavy elements. In stellar interiors, the heavier elements of the carbon group and the metals primarily determine the opacity.

open cluster (sometimes called galactic cluster) A comparatively loose grouping (mass range 10^2–10^3 $M\odot$) of Population I stars, strongly concentrated in the spiral arms or the disk of the Galaxy (in fact, open clusters give a good indication of where the spiral arms are). Unlike associations, open clusters are dynamically stable. Depending on their age, stars in open clusters "peel off" from the main sequence at different points (the higher the turnoff point, the younger the cluster). (NGC 188 is the oldest known open cluster.)

open universe A big-bang model of the Universe with a hyperbolic geometry—i.e., one in which the initial velocity of the particles in the big bang exceeded the "escape velocity" of the universe.

70 Oph A visual binary (K0 V, K5 V) 4.9 pc distant (1974 π = $0\rlap{.}''203$). Period 88.13 years.

ζ Ophiuchi A reddened O9.5 V star 170 pc distant (a runaway star from the Sco-Cen association) with a high rotational velocity (396 km s^{-1}). It is well known for its strong interstellar absorption lines in the visible part of the spectrum.

Oppenheimer-Volkoff limit The limiting mass for a neutron star as the density approaches infinity. Beyond this mass all configurations are unstable.

opposition See elongation.

optical depth (τ) A measure of the integrated opacity along a path through a layer of material, measured by the amount of absorption of a beam of incident light. The intensity ratio $I/I_0 = e^{-\tau}$, where $\tau = N\sigma l$ (N = column density, σ = cross section, and l is the path length). Optical depth 1 corresponds to the "visible" surface and occurs when the intensity is reduced by a factor e.

optical pair A pair of stars that appear close together on the sky as a result of perspective only, and that have no physical relation.

optical window A gap in Earth's atmospheric absorption spectrum through which visible light can pass down to the surface. The optical window includes the spectral region between the O_3 cutoff at 2950 Å and the A band of O_2 at 7600 Å.

orbital elements Seven quantities needed to establish the orbit of a celestial body (see elements of an orbit).

orbital velocity Velocity required by a body to achieve a circular orbit around its primary: $V_{orb} = (GM/r)^{1/2}$.

orbiting collision A "collision" in which an ion and an atom approach each other very closely and spend a long time (several orbits of the atomic electrons) in close proximity.

Orgueil meteorite   A Type I carbonaceous chondritic meteorite that fell in France in 1864 and that has recently been found to contain amino acids.

Orion A (3C 145)   A radio continuum feature (an H II region) centered on the Trapezium, and excited by θ^1 Ori C. The Orion A molecular cloud, which lies beyond it, is a rich source of molecules—CO, OH, HCN, and probably NO, HCO, and H_2CO have been observed.

Orion arm (also called local arm)   The spiral arm of the Milky Way on a spur of which the Sun is located (see Orion spur). It is about 600 pc across, and lies about 10.4 kpc from the galactic center, between the Sagittarius arm and the Perseus arm. The total density of interstellar gas in the Orion arm is about 1.5 atoms cm^{-3} (density of H I about 0.6 atoms cm^{-3}).

Orion B   A radio continuum source (NGC 2024).

Orion Molecular Cloud 1 (OMC-1)   Centered approximately $1'$ northwest of the Trapezium, it contains the Becklin-Neugebauer and Kleinmann-Low infrared sources.

Orion Molecular Cloud 2 (OMC-2)   An infrared and molecular emission complex about $12'$ northeast of the Trapezium, centered on a cluster of infrared sources.

Orion Nebula (M42, NGC 1976)   An H II region about 500 pc distant, barely visible to the naked eye in the center of Orion's sword. It is undoubtedly a region where stars are being born; young O stars and many T Tauri variables are associated with it, and its members are extreme Population I. Probably no more than 20,000 years old. It is also an X-ray source (3U 0527-05).

Orion spur   That part of the local spiral arm in which the Sun is embedded. (The Sun is on an inner edge of the Orion spur.)

α Orionis   See Betelgeuse.

β Orionis   See Rigel.

BM Orionis   A peculiar eclipsing binary (B2–B3) in the Trapezium, with a flat-bottomed light curve suggesting a total eclipse. The spectrum of the secondary has never been seen.

FU Orionis   A newly formed star, probably a pre–main-sequence star (cF5–G3 Ia) presently near the top of its Hayashi track. In 1936 it suddenly appeared in the middle of a dark cloud, and rose by 6 magnitudes in the photographic band. Its lithium abundance is 80 times that of the Sun. It has developed a reflection nebula.

θ^1 Orionis   See Trapezium.

θ^2 **Orionis**   A 21.03-day O9.5 Vp spectroscopic binary tentatively identified with 2U 0525 − 06.

YY Orionis   An extremely young star (younger than T Tauri) in the Orion Nebula. YY Orionis stars are very young, late-type, low-mass stars in the gravitationally contracting stage in which the star is still accreting matter from the protostellar cloud.

ortho-hydrogen   Molecular hydrogen in which the two protons of the diatomic molecule have the same direction of spin. It is a higher energy state than the para form. Terrestrial H_2 is 75% ortho-hydrogen, 25% para-hydrogen.

orthonormal tetrad   A set of four mutually orthogonal unit vectors at a point in spacetime, one timelike and three spacelike, which give the directions of the four axes of a locally Minkowskian coordinate system.

ortho-spectrum   Spectrum of triplet ($l = 1$).

oscillator strength (*f*-value)   A measure of the probability that a transition represented by an electronic oscillator will occur. It is independent of the physical conditions under which the atom is radiating.

oscillating universe   A version of the big-bang theory which has a spherical geometry and in which the expansion curve is a cycloid. In this model the universe continuously undergoes successive cycles of expansion and collapse.

osculatory elements   Orbital elements used in calculating perturbations.

osculating orbit   The path that an orbital body (e.g., a planet) would follow if it were subject only to the inverse-square attraction of the Sun or other central body. In practice, secondary bodies, such as Jupiter, produce perturbations.

outgassing   Ejection of the gases locked in the interior of a planet so that they become part of the planet's atmosphere.

overshoot   A condition that obtains when the momentum of a particle carries it past its equilibrium point.

overstability   A form of instability which, when it sets in, sets in as oscillations of increasing amplitude.

overtone   See harmonic overtone.

"Owl" Nebula   A planetary nebula (M97, NGC 3587) in Ursa Major, ∼600 pc distant.

ozone layer   A layer in the lower part of Earth's stratosphere (about 20–60 km above sea level) where the greatest concentration of ozone (O_3) appears. This is the layer responsible for the absorption of ultraviolet radiation.

P

P-branch   A set of lines in the spectra of molecules corresponding to unit increases in rotational energy.

p-electron   An orbital electron whose l quantum number is 1.

pep reaction   A reaction occurring in the proton-proton chain. The first step, instead of $p + p \rightarrow d + e^+ + \nu_e$, is $p + e^- + p \rightarrow d + \nu_e$. This latter reaction occurs only once in 400 p-p reactions but produces far more energetic neutrinos (1.44 MeV as against 0.42 MeV).

p-p chain   See proton-proton chain.

p-process   The name of the hypothetical nucleosynthetic process thought to be responsible for the synthesis of the rare heavy proton-rich nuclei which are bypassed by the r- and s-processes. It is manifestly less efficient (and therefore rarer) than the s- or r-process, since protons must overcome the Coulomb barrier, and may in fact work as a secondary process on the r- and s-process nuclei. It seems to involve primarily (p,γ) reactions below cerium (where neutron separation energies are high) and the (γ,n) reactions above cerium (where neutron separation energies are low). The p-process is assumed to occur in supernova envelopes at a temperature greater than about 10^9 K and at densities less than about 10^4 g cm^{-3}.

p-spot   See sunspot.

p-wave (the p stands for primary)   A longitudinal seismic acoustic wave that moves by compression. The p-waves travel faster than s-waves and can penetrate the core of the Earth.

packing fraction   Mass defect per nuclear particle. The term has been largely superseded by the related quantity, binding energy per nuclear particle.

pair annihilation   Mutual annihilation of an electron-positron pair with the formation of gamma rays, or of a proton-antiproton pair with the formation of pions. The charges cancel, and the total mass of the pair is converted into energy (unlike nuclear fusion, in which less than 1% of the mass is converted into energy).

pair production   The inverse process to pair annihilation. A gamma ray of the right energy (> 1.02 MeV) is transformed into an electron-positron pair (or a pion is transformed into a proton-antiproton pair).

pairing energy (δ)   A quantity which expresses the fact that nuclei with odd numbers of neutrons and protons have less energy and are less stable than those with even numbers of neutrons and protons.

Pallas   The second asteroid to be discovered (by Olbers in 1802). Diameter about 560 km; $a = 2.77$ AU, $e = 0.235$, $i = 34°8$. Orbital period 1,686 days; rotation period 9–12 hours. Albedo ∼0.05; mass (1972 est.) 2.6×10^{23} g. Spectrum resembles meteorites of either low-grade carbonaceous chondrite or enstatite achondrite.

Pan   Unofficial name for Jupiter XI. $P = 692^d$ R, $e = 0.2$; $i = 163°$. Discovered by Nicholson in 1938.

para-hydrogen   Molecular hydrogen in which the two protons of the diatomic molecule have opposite directions of spin. It is a lower energy state than ortho-hydrogen.

parallax (stellar)   The angle subtended by 1 AU at the distance of a nearby star. Stellar parallaxes can be measured down to about $0°01$, which corresponds to a distance of 100 pc (D[pc] = $1/\pi$[arc sec]), but are valid (to $\leq 10\%$) only to about 20 pc. First trigonometric parallax was obtained in 1838.

parametric amplifier (paramp)   A device used in radio astronomy for increasing the strength of a radio signal.

parasites   In radio jargon, spiral coils or gratings of wire used on dipole antennas of radio telescopes to give greater sensitivity.

para-spectrum   Spectrum of singlet ($l = 0$).

parity   The principle of space-inversion invariance; i.e., no experiment can differentiate between the behavior of a system and that of its mirror image. Parity is conserved in strong interactions, but not in weak ones.

parsec (abbreviation for parallax second)   The distance at which one astronomical unit subtends an angle of 1 second of arc. 1 pc = 206,265 AU = 3.086×10^{13} km = 3.26 lt-yr.

particle distribution function   The number of particles per unit volume of phase space (q.v.).

partition function   The effective statistical weight of an atom or ion under existing conditions of excitation or ionization.

parton   A hypothetical pointlike constituent of a nucleon, which contains all the charge of the nucleon.

pascal (Pa)   The derived SI unit of pressure. 1 Pa = 1 N m^{-2} = 10^{-5} bars.

Paschen-Back effect   An effect on spectral lines obtained when the light source is located in a strong magnetic field, so that the magnetic splitting becomes greater than the multiplet splitting.

Paschen series   The spectral series associated with the third energy level of the hydrogen atom. The series lies in the infrared—Pα at 18,751 Å; Paschen limit at 8204 Å. (Pα of He II is at 4686 Å; He II series limit is at 2040 Å.)

passband   The frequency band that is transmitted with maximum efficiency and without intentional loss.

past light cone   See light cone.

Patroclus   Asteroid 617, a Trojan 60° behind Jupiter. $P = 11.82$ yr, $a = 5.19$ AU, $e = 0.14$, $i = 22°.1$.

Paul trap   A radiofrequency quadrupole ion trap in which charged particles can be suspended by radiofrequency electric fields for times limited primarily by collisions with the background gas.

Pauli exclusion principle   See exclusion principle.

pc   Abbreviation for parsec (q.v.).

peculiar stars   Stars with spectra that cannot be conveniently fitted into any of the standard spectral classifications. They are denoted by a p after their spectral type.

peculiar velocity   Velocity with respect to the local standard of rest (q.v.).

pencil beam   The main lobe of an antenna pattern, consisting of a narrow receiving beam of the type obtained with a single parabolic instrument.

Penrose process   A means of extracting energy from a rotating black hole. If a particle spirals into the ergosphere of a black hole in a direction counter to the rotation of the black hole, and if the particle then breaks up into two fragments inside the ergosphere, one of the fragments can escape with energy greater than the energy of the original particle.

Penrose's theorem   A collapsing object whose radius is less than its gravitational radius must collapse into a singularity.

perfect cosmological principle   The assumption adopted by the steady-state theory, that all observers, everywhere in space and at all times, would view the same large-scale picture of the Universe.

perfect gas   See ideal gas.

periapsis   The point in the orbit of a satellite where it is closest to its primary.

periastron   The point in the orbit of one component of a binary system where it is nearest the other component.

pericenter   The point in the orbit of one component of a binary system which is closest to the center of mass of the system.

pericynthion   The point in the orbit of a satellite around the Moon closest to the Moon.

perigalacticon   The point in the orbit of a star which is nearest the center of the Galaxy.

perigee   The point in the orbit of an Earth satellite where it is closest to the Earth's center of mass.

perihelion   The point in the orbit of an object orbiting the Sun where it is closest to the Sun's center of mass. Earth's perihelion occurs early in January.

period-luminosity relation   A correlation between the periods and mean luminosities of Cepheids, discovered by Henrietta Leavitt in 1912. It is an important distance indicator out to about 3 Mpc.

α **Persei**   A young open cluster with a high mean rotational velocity.

β **Persei stars**   A class of eclipsing binaries (see Algol) with periods of from 2 to 5 days, the depth of whose secondary minimum is almost negligible.

h and χ Persei (also called Perseus OB1)   A double stellar association about 2 kpc distant, visible to the naked eye as a patch of light. It contains many young O and B stars and also many M supergiants.

Perseus A (3C 84, Abell 426)   A strong radio source ($z = 0.0183$; recession velocity about 5000 km s^{-1}). Optically it is a Seyfert galaxy (NGC 1275) with a huge amount (about 10^8 $M\odot$) of ionized gas receding from it. It is also a strong X-ray source (3U 0316+41).

Perseus arm   A spiral arm of the Milky Way located in the direction of Perseus about 12.3 kpc from the galactic center.

Perseus cluster   A diffuse, irregular cluster of about 500 galaxies ($z = 0.0183$) (richness class 2) dominated by and centered on the Seyfert galaxy NGC 1275 (see Perseus A). Mass required to bind the cluster, greater than 10^{15} $M\odot$; mass of cluster, about 2×10^{15} $M\odot$.

Perseus OB1   See h and χ Persei.

Perseus OB2 (Perseus 2)   A young association of OB stars about 350 pc distant.

Perseus X-1 (3U 0316+41)   The strongest known extragalactic X-ray source, centered on NGC 1275.

perturbation   A small disturbance which makes the system deviate from its equilibrium state. It is by considering such perturbations that one determines the stability of a system: it is stable if in time the system returns to its equilibrium state; and it is unstable if some initial perturbation makes the system depart from the equilibrium state indefinitely.

perturbation method   A system of successive approximations to the solution of a problem, by starting with a closely similar problem whose solution is known, applying small departures from equilibrium, and then calculating their consequences.

Pfund series   A spectral series of hydrogen lines in the far-infrared, representing transitions between the fifth energy level and higher levels.

phase   The fraction of the period between the instants at which two oscillations or waves attain their maximum displacement. If the maxima coincide, the two waves are said to be in phase. Also, the name given to the changing shape of the visible illuminated surface of a non–self-luminous celestial body (the Moon or a planet). The phase changes are caused by the relative positions of the Earth, Sun, and illuminated body. Conventionally, 0° phase occurs when the hemisphere facing the Earth is fully sunlit.

phase angle   The difference in phase between two waves. See also solar phase angle.

phase space   A six-dimensional mathematical space that includes not only the three dimensions of ordinary space but also the three dimensions of velocity space (q.v.). A point in phase space represents a given position in ordinary space and a given velocity in velocity space.

phase switching   A technique used in radio astronomy to suppress background noise so that the receiver records only point sources.

Phillips bands   Spectral bands of the C_2 molecule in the red and near-infrared (0–0 band at 1.207 μ).

Phobos   The potato-shaped inner satellite of Mars (about 18 × 22 km), discovered by A. Hall in 1877. Orbital and rotation period $7^h39^m14^s$, $e = 0.021$, $i = 1°.1$. Visual geometric albedo 0.06. Infrared observations suggest that its surface is covered with dust. Phobos lies just outside the Martian Roche limit.

Phoebe   The outermost satellite of Saturn, discovered by Pickering in 1898. Period 550 days retrograde; radius about 100 km.

SX Phoenicis   A dwarf Cepheid (spectral type A) with the shortest known period (1^h19^m).

phonon   The quantum associated with lattice vibrations in a solid. Phonons are sound quanta.

photoelectric magnitude (m_{pe})   The magnitude of an object as measured with a photoelectric photometer.

photographic magnitude (m_{ph})   The magnitude of an object as measured on the traditional photographic emulsions, which are sensitive to a slightly bluer region of the spectrum than is the human eye.

photoionization   The ionization of an atom or molecule by the absorption of a high-energy photon by the particle. It is an important source of opacity in stars.

photometric binaries   Eclipsing variables like β Per whose orbital plane lies so nearly in the line of sight that eclipses, as seen from the Earth, can occur and can be detected from their light curves.

photometry   The measurement of light intensities by taking a picture of a star field and then measuring the intensities of the stars.

photomultiplier tube   A photoelectric cell which uses the principle of the photoelectric effect. Photons of incident starlight free electrons from a sensitive surface, and the electric current thus generated is amplified in about 10 stages within the tube. It is so sensitive that it can detect the incidence of a single photon.

photon   A discrete quantity of electromagnetic energy; a quantum of light (see boson). Photons have spins of ± 1, a rest mass of zero, and are their own antiparticles.

photoneutrinos   Neutrino-antineutrino pairs produced by the collision of high-energy photons with electrons: $\gamma + e^- \rightarrow e^- + \nu + \bar{\nu}$.

photosphere   The region of a star which gives rise to the continuum radiation emitted by the star. The visible surface of the Sun (temperature about 6000 K), just below the chromosphere and just above the convective zone. The photosphere ends (and the chromosphere begins) at about the place where the density of negative hydrogen ions has dropped to too low a value to result in appreciable opacity. The spectrum of the photosphere consists of absorption lines (unlike that of the chromosphere, which consists of emission lines).

photovisual magnitude   The magnitude of an object as measured photographically by filters and emulsions that are sensitive to the same region of the spectrum as the human eye.

Pickering series   A spectral series of He II lines found in very hot O-type stars. It is associated with the fourth energy level—Pi α at 10124 Å; Pi β at 6560 Å. The series limit is at 3644 Å.

pico-   A prefix meaning 10^{-12}.

pinch machine   A fusion device containing a plasma heated by a shock wave generated within the plasma as it is constricted by the rapidly increasing magnetic field.

pion (also called π-meson)   An unstable nuclear particle of mass intermediate between that of a proton and an electron (π^+ and π^-:
273 m_e; π^0: 264 m_e). The pions are believed to be the particles
exchanged by nucleons, resulting in the strong nuclear force;
they play a role in the strong interactions analogous to that of
the photons in electromagnetic interactions. A charged pion
usually decays into a muon and a neutrino; a neutral pion, into
two γ-rays. Pions have spins of 0.

pitch angle   Angle specifying the direction of electron velocity; or the
angle between a tangent to a spiral arm and the perpendicular to
the direction of the galactic center.

pixel   Shortened term for "picture element." It is a resolution element
in a vidicon-type detector.

P-L relation   See period-luminosity relation.

plage (sometimes called flocculus)   The bright rim of a sunspot, observed in emission in monochromatic light of some spectral line
(Hα or Ca II). It is a chromospheric phenomenon associated
with and often confused with a facula.

Planck's blackbody formula   A formula that determines the distribution of intensity of radiation that prevails under conditions of
thermal equilibrium at a temperature T: $B_\nu = (2h\nu^3/c^2)[\exp(h\nu/kT) - 1]^{-1}$ where h is Planck's constant and ν is the frequency.

Planck's constant (h)   The constant of proportionality relating the frequency of a photon to its quantum of energy: $h \approx 6.626 \times 10^{-27}$
erg seconds.

Planck length $[(G\hbar/c^3)^{1/2} = 1.6 \times 10^{-33}$ cm] The dimension at which
space is predicted to become "foamlike" and at which Einstein's
theory is supposed to break down.

plane-parallel atmosphere   An atmosphere stratified in parallel planes
normal to the direction of gravity.

planetary nebula   An expanding envelope of rarefied ionized gas surrounding a hot white dwarf. The envelope receives ultraviolet
radiation from the central star and reemits it as visible light by
the process of fluorescence. The planetary nebula stage lasts for
less than 50,000 years. During the core contraction that termi-
nates the red-giant stage, the helium-burning shell is ejected at a
velocity so high that it becomes separated from the core. Under
current theories, a star with a carbon core and a mass greater
than 0.6 $M \odot$ (but less than 4 $M \odot$) will become a planetary
nebula and leave behind a white dwarf. Planetary nebulae are
now known to occur in stars less than 4 $M \odot$ whose envelope

becomes unstable during the hydrogen shell burning stage.

planetary precession   The component of general precession (q.v.) caused by the gravitational coupling between the center of mass of the Earth and that of the other planets. The effect of planetary precession is to move the equinox *eastward* by ~ 0.11/year and to diminish the angle between the ecliptic and the equator by about 0.47/year.

Plaskett's star (HD 47129)   A very massive O-type giant with known anomalies in its spectrum. It is a spectroscopic binary in which mass exchange is occurring. Its spectrum can be interpreted to mean that each component has a mass of 75 $M\odot$.

plasma   A completely ionized gas; the so-called fourth state of matter (besides solid, liquid, and gas) in which the temperature is too high for atoms as such to exist and which consists of free electrons and free atomic nuclei.

plasma clouds   Clouds of electrically charged particles embedded in the solar wind.

plasmapause   The region in Earth's ionosphere (at about 4–7 Earth radii) where the particle density (100 particles per cm^3 just below the plasmapause) drops off very rapidly. It marks the transition from high to low density.

Pleiades (M45, NGC 1432)   A very young open cluster of several hundred stars (B6 and later) in Taurus, about 125 pc distant. Six members of the cluster (all of spectral type B or Be) are visible to ordinary sight.

Pleione   A B8pe star (28 Tau), one of the brightest stars in the Pleiades, which developed an envelope or shell first observed in 1938. The shell increased in strength and attained its maximum intensity in 1945; thereafter it weakened and was scarcely visible by 1954. In 1972 it developed another shell. It is rotating so fast that it is unstable.

plume   A rising column of gas over a maintained source of heat.

Pluto   The most distant known planet from the Sun (39.44 AU), discovered by Tombaugh in 1930. Orbital period 248.43 years, V_{orb} 4.7 km s^{-1}. Its orbit has the highest eccentricity (0.249) and highest inclination to the ecliptic (17.17) of any planet, and some astronomers suggest that it may be an escaped satellite of Neptune. Synodic period 366.7 days; albedo less than 0.25; rotation period $6^d9^h17^m49^s$. In the mid-1970s Pluto crosses Neptune's orbit on its way in, and for the rest of this century Pluto will be closer to the Sun than Neptune. (Pluto and Neptune, however, are *never* less than 2.6 AU apart.) Perihelion will occur in 1989.

Effective temperature about 50–60 K. Its mass and radius have not been determined with any great certainty, but it is probably about 0.1 to 0.2 the mass of the Earth (6×10^{26} g?) and no more than 2900 km in radius.

Pockels cell   An electro-optic crystal used as a reversible waveplate by applying alternately high positive and negative voltage.

Pogson's ratio   The ratio between two successive stellar magnitudes (q.v.), introduced by N. Pogson in 1856.

Poincaré's theorem   The total kinetic energy of all the stars in a cluster is equal to half the negative gravitational potential energy of the cluster.

point source   A source whose angular extent cannot be measured ($< 0''.05$).

Poisson distribution   An approximation to the binomial distribution used when the probability of success in a single trial is very small and the number of trials is very large.

Poisson's equation   An equation ($\nabla^2 \phi = 4\pi G\rho$) which relates the gravitational (or electromagnetic) potential to the mass density (or charge density).

Polaris (α UMi)   A supergiant F8 Ib, F3 V visual binary, 120 pc distant, with an orbital period of thousands of years. The primary (a Cepheid with a pulsation period of 3.97 days) is itself a single-lined spectroscopic double with a period of 29.6 years. There are at least two more faint (12th mag) components of the system.

Pollux (β Gem)   A K0 III star 11 pc distant.

polytrope   A mathematical model of an inhomogeneous, compressible configuration in equilibrium under its own gravitation in which the relation between the pressure and the density satisfies the relation $p = K\rho^{(n+1)/n}$, where K is a constant and n is the polytropic index.

polytropic index (n)   See polytrope. The polytropic index may have any value from zero (uniform density throughout) to 5 (entire mass concentrated at the center). A polytropic index of 1.5 corresponds to a fully degenerate, nonrelativistic electron gas; it also describes a perfect-gas star in convective equilibrium.

population inversion   A condition that exists when there are more molecules in an excited state than an equilibrium distribution would allow. It is necessary for masers.

Populations I and II   Two classes of stars introduced by Baade in 1944. In general, Population I (now sometimes called arm population) are young stars with relatively high abundances of metals, and are usually found in the disk of a galaxy, especially the spiral arms, in dense regions of interstellar gas. Population II (now sometimes called halo population) are older stars with relatively low abundances of metals, and are usually found in the nucleus of a galaxy or in globular clusters. The Sun is a rather old Population I star.

Poseidon   Unofficial name for J VIII, the next outermost satellite of Jupiter. $P = 737^dR$, $e = 0.4$, $i = 147°$. Discovered by Melotte in 1908.

position angle   Angular distance (in degrees, measured from north through east) between the primary and secondary components of a binary system.

positron (also called antielectron)   A particle with the mass of an electron but with an equal positive charge. It is the antiparticle with respect to the electron.

post-Galilean transformation   A transformation which replaces the Lorentz transformation when first-order corrections due to general relativity are included.

post-Newtonian effects   The first nontrivial gravitational effects which go beyond the predictions of Newton's theory.

power series   A series of the form $a_0 + a_1v + a_2v^2 + \ldots + a_nv^n = \sum_{n=1}^{\infty} a_nv^n$.

Poynting-Robertson effect   An effect of radiation pressure on a small particle orbiting the Sun that causes it to spiral slowly into the Sun. The radiation falls preferentially on the leading edge of the orbiting particle and acts as a drag force.

Praesepe (Beehive Cluster) (M44, NGC 2632)   A naked-eye open cluster in Cancer, about 160 pc distant.

Prandtl number   Ratio of the product of the viscosity coefficient and the specific heat at constant pressure to the thermal conductivity.

precession   A slow, periodic conical motion of the rotation axis of a spinning body. In the case of Earth's precession it is due to the fact that Earth's axis of rotation is not perpendicular to the ecliptic but is inclined about $23°.5$ and is thus affected by gravitational perturbations from other bodies in the solar system. The Moon and Sun pull harder on that part of the Earth's equatorial bulge nearest them than on that farthest away; this causes a torque which precesses the Earth's rotational axis.

precession, constant of   The ratio of the lunisolar precession to the cosine of the obliquity of the ecliptic. It amounts to about $54''.94$ per annum.

precession of the equinoxes   The slow westward drift (50."26 per year) of the equinoxes due to Earth's precession (q.v.). Because of precession the tropical year is about 20 minutes shorter than the sidereal year. It takes about 25,800 years for Earth's axis to complete one circuit.

precursor pulse   A component of a pulsar pulse which occurs slightly before the main pulse. At energies of about 600 keV the precursor pulse becomes stronger than the main pulse.

pressure   The force exerted over a surface divided by its area.

pressure broadening   A broadening of spectral lines, particularly in white dwarfs, caused by the pressure of the stellar atmosphere, which in turn is caused by the surface gravity of the star.

pressure ionization   A state found in white dwarfs and other degenerate matter in which the atoms are packed so tightly that the electron orbits encroach on each other to the point where an electron can no longer be regarded as belonging to any particular nucleus and must be considered free.

pressure scale height   See scale height.

primary cosmic rays   The cosmic rays (q.v.) that arrive at Earth's upper atmosphere from outer space (cf. secondary cosmic rays).

primeval fireball   The primeval "atom," containing all the mass of the universe, whose explosion was responsible, according to the big-bang theory, for the present expanding universe. At 10^{-43} seconds after the big bang, when the density of the universe was 10^{93} g cm^{-3}, the temperature had dropped to 10^{12} K.

principal quantum number (n)   A measure of the major axis of an electronic orbital. In the case of hydrogen, the energies of bound levels are specified completely by n.

probable error (p.e.)   The error which will not be exceeded by 50 percent of the cases. The probable error is equal to 0.6745 times the standard error.

Procyon (α CMi)   An F5 IV–V star 3.5 pc distant (parallax 0.283). It is a visual binary; its companion is a DF8 white dwarf with an orbital period of ∼40 yr.

profile   See line profile.

program stars   The stars being observed or measured, as contrasted with the comparison stars.

prograde motion   Motion in the same direction as the prevailing direction of motion.

promethium (Pm)   An unstable rare earth. The longest-lived isotope, ^{145}Pm, has a half-life of only 18 years.

prominence   A region of cool (10^4 K), high-density gas embedded in the hot (10^6 K), low-density solar corona. Prominences are the flamelike tongues of gas that appear above the limb of the Sun.

proper mass   Rest mass.

proper motion   Apparent angular rate of motion of a star across the line of sight on the celestial sphere.

proper time   The timelike invariant spacetime interval between the points along the trajectory of a particle. (More prosaically, time measured by an ideal clock at rest with respect to the observer.)

proportional counter   A device used in X-ray astronomy which counts the number of ions produced when photons come into a volume of gas and ionize the gas. The more energetic the photon, the more ions are produced.

proton   A positively charged elementary particle; the nucleus of a hydrogen atom. Mass of proton 1.00728 amu $= 1.6726 \times 10^{-24}$ g $= 1836.12\ m_e$.

proton-proton chain (*p-p* chain)   A series of thermonuclear reactions in which hydrogen nuclei are transformed into helium nuclei. The temperature and density required are about 10^7 K and 100 g cm^{-3}. It is the main source of energy in the Sun, where 10^{38} of these reactions occur every second. All parts of this reaction have been observed in the laboratory, except for the first step ^{1}H($p,\beta^+\nu$)^{2}D, which occurs only a few times in 10^{12} collisions of protons. But the first two reactions provide about one-third of the Sun's total energy release. The *p-p* chain divides into three main branches: PP I: ^{1}H($p,\ \beta^+\nu$)^{2}D $(p,\gamma)^3$He(^{3}He, $2p)^4$He $+ 4 \times 10^{-5}$ ergs of energy. PP II: ^{1}H($p,\beta^+\nu$)^{2}D(p,γ)^{3}He (^{4}He, γ) ^{7}Be($\beta^+\nu$)^{7}Li(p,α)^{4}He. PP III: ^{1}H($p,\ \beta^+\nu$)^{2}D($p,\ \gamma$)^{3}He(^{4}He, γ)^{7}Be(p,γ)^{8}B ($\beta^+\nu$)^{8}Be $\rightarrow$ 2 ^{4}He. (PP III occurs once in 1000 times.) Although the neutrinos from the PP II and PP III chains are detectable, they have not been observed.

pulsar   An object discovered at Cambridge University in 1967 which has the mass of a star and a radius no larger than that of Earth and which emits radio pulses with a very high degree of regularity (periods range from 0.03 s for the youngest to more than 3 s for the oldest). All pulsars are characterized by the general properties of dispersion, periodicity, and short duty cycle. Pulsars are believed to be rotating, magnetic (surface magnetic fields of 10^{10} to 10^{14} gauss are estimated) neutron stars which are the end products of supernovae. Type S pulsars have a simple pulse shape; Type C, complex; Type D have drifting subpulses.

pulse counter   When an atom is ionized by collision with a charged particle, the electrons it loses can be collected by applying a voltage. The process of collection gives an electrical pulse that is proportional to the number of free electrons, which in turn is proportional to the energy of the colliding particle.

pulse counting   Counting each individual photon as it comes off.

pulse width   The interval of time between two successive pulses.

pulse window   See mean profile.

pumping (optical)   A process of raising matter from lower to higher energy levels. In order for a maser to work continuously, there must be some mechanism that replenishes the energy depleted by the emission and that provides population inversion. Such a mechanism is known as a pump.

Puppis A   A supernova remnant 10^4–10^5 years old, about 1–2 kpc distant. It is an extended nonthermal radio source, and also a source of soft X-rays (2U 0821 − 42).

ζ Pup   An extremely bright O4f star (the brightest Of star known) embedded in the Gum Nebula. It has an envelope which is rapidly accelerating outward.

pycnonuclear   An adjective used to describe nuclear processes (such as the proton-proton chain) that take place at relatively low temperatures and that are not strongly temperature-dependent.

Q

q_0   See deceleration parameter.

QSO   See quasar.

Q-branch   A set of lines in the spectra of molecules corresponding to changes in vibrational energy with none in rotational energy.

quadrature   Elongation (q.v.) of a planet when it makes a 90° angle with the Sun as seen from Earth. Or, a calculation involving a definite integral.

quadrupole   When referred to a system containing charges, a quadrupole is equivalent to the presence of two equal dipoles parallel to each other, but with their corresponding charges reversed; or more generally, that component of the charge distribution which has axial or triaxial symmetry. Similarly, when referred to mass distributions, it arises from unequal components of the moment-of-inertia tensor along three principal directions.

quantization   The restriction of various quantities to certain discrete values; or, more generally, to deriving the quantum-mechanical laws of a system from its corresponding classical laws.

quantum   A discrete quantity of energy $h\nu$ associated with a wave of frequency ν. It is the smallest amount of energy that can be absorbed or radiated by matter at that frequency.

quantum defect (also called Rydberg correction)   The principal quantum number responsible for a spectral series, minus the Rydberg denominator for any actual spectral term of the series.

quantum efficiency   The efficiency of a counter in detecting photons; the probability that a photon will liberate an electron and thus be detected.

quantum field theory   The relativistically invariant version of quantum mechanics.

quantum solid   A degenerate gas in which the densities are so great that the nuclei are fixed with respect to each other so that they resemble a crystalline lattice.

quantum theory   Initially, the theory developed by Planck that radiating bodies emit energy not in a continuous stream, but in discrete units called quanta, the energy of which is directly proportional to the frequency. Now, all aspects of quantum mechanics.

quantum yield (in photochemistry)   Number of molecules decomposed per photon absorbed.

quarks   A hypothetical group of elementary particles which may be combined to produce the baryons and mesons. They are of three types, p, n, and λ, each with a fractional charge and each having an antiparticle, and all having half-integral spins. According to hypothesis, each baryon is a multiplet composed of three quarks and each meson is a doublet composed of two quarks (a quark and an antiquark).

quasar (also called quasi-stellar object or QSO)   An object with a dominant starlike (i.e., diameter less than $1''$) component, with an emission line spectrum showing a large redshift—up to $z = 3.53$ $(0.91c)$ for OQ 172. (The largest redshift known for a normal radio galaxy is $z = 0.637$ for 3C 123.) Many have multiple absorption redshifts; a few have multiple emission redshifts. (Bahcall system: class I, $z_{abs} \approx z_{em}$; class II, z_{abs} significantly less than z_{em}.) The light of most if not all quasars is variable over time intervals between a few days and several years, so their diameters must not be much larger than the diameter of the solar system; yet they are the intrinsically brightest objects known (for 3C 273 (z = 0.158), M_V = −27.5 if its redshift is cosmological). The energy output of a typical quasar at "cosmological" distance is of the order of 10^{47} ergs per second—which would require a mass of 10^{10} $M\odot$ if it derives its energy solely from nuclear fusion. (Energy requirement under the "local" hypothesis is on the order of 10^{42} ergs per second.) The basic problem of quasars is that they emit too much radiation in too short a time from too small an area.

quasi-stellar radio source   A quasar (q.v.) with detectable radio emission.

quiet Sun   The Sun when the 11-year cycle of activity is at a minimum.

R

R-branch   A set of lines in the spectra of molecules corresponding to unit decreases in rotational energy.

R galaxy   In the Yerkes 1974 system, a system showing rotational symmetry, without clearly marked spiral or elliptical structure (formerly called D galaxy).

r-process   The capture of neutrons on a very rapid time scale (i.e., one in which a nucleus can absorb neutrons in rapid succession, so that regions of great nuclear instability are bridged), a theory advanced to account for the existence of *all* elements heavier than bismuth (up to $A \approx 298$) as well as the neutron-rich isotopes heavier than iron. The essential feature of the *r*-process is the release of great numbers of neutrons in a very short time (less than 100 seconds). The presumed source for such a large flux of neutrons is a supernova, at the boundary between the collapsing neutron star and the ejected material. However, other proposed sources have included such things as supernova shocks and black-hole-neutron-star collisions. The heavier *r*-process elements are synthesized at a temperature of about 10^9 K and an assumed neutron density of $10^{20} - 10^{30}$ per cm^3. The *r*-process is terminated by neutron-induced fission. The existence of ^{244}Pu (half-life 82 million years) in the early solar system shows that at least one *r*-process event had occurred in the Galaxy just before the formation of the solar system.

R star   Stars of spectral type R are stars with spectral characteristics similar to those of K stars except that molecular bands of C_2,

CN, and CH are present instead of TiO bands.

R zones   Regions in the solar corona in which short-lived radiofrequency variations are observed.

rad   Unit of radiation, equal to 100 ergs of ionizing energy absorbed per gram of absorber.

radial velocity   Velocity along the line of sight toward $(-)$ or away from $(+)$ the observer.

radiant (or vertex)   The convergent point toward which the stars in a moving cluster appear to travel, or from which the meteors in a shower seem to radiate.

radiation damping   A decrease in the amplitude of an oscillation due to the emission of energy by radiation.

radiation era   The era from about 10 s to about 10^{12} s after the big bang, when the temperature had dropped to 10^9 K and the rate of electron-positron pair annihilation exceeded the rate of their production, leaving radiation the dominant constituent of the universe. At $t = 200$ s, nucleosynthesis began rather abruptly and virtually all deuterium was synthesized to helium. The radiation era was followed by the matter era (q.v.).

radiation length   The mean distance traveled by a relativistic particle in a given medium before its energy is reduced by a factor e by its interaction with matter.

radiation pressure   The transfer of momentum by electromagnetic radiation incident on a surface: $p_{rad} = (4/3)\sigma T^4/c$.

radiation temperature   The temperature that a blackbody of similar dimensions would have that radiated the same intensity at the same frequency.

radiative braking   The slowing down of rotation of a star due to radiation.

radiative capture   Capture of a free electron by an ion with the subsequent emission of an X-ray (or gamma-ray) photon (also called radiative recombination).

radiative recombination   See radiative capture.

radioactivity   The spontaneous disintegration of unstable atomic nuclei. All natural radioactive elements heavier than lead are daughter products of either ^{232}Th (half-life 1.39×10^{10} yr), ^{235}U (half-life 7.13×10^8 yr), or ^{238}U (half-life 4.51×10^9 yr). The radioactive output of the Earth averages 1.7 ergs g^{-1} yr^{-1}.

radio astronomy   The branch of astronomy that makes observations at radio wavelengths. The observable in radio astronomy is the difference between the amount of radiation received with the telescope pointed at the source and the amount received with the telescope pointed at a baseline position slightly off the source.

radio galaxy   A galaxy that is extremely luminous at radio wavelengths. A radio galaxy is usually a giant elliptical—the largest galaxy in a cluster—and is a strong emitter of synchrotron radiation. M87 and M82 are examples.

radio recombination lines   See spectral lines. Radio recombination lines are the result of electron transitions between high-n ($n >$ 50) levels in an atom or ion.

radiosonde   A sounding balloon used to transmit information on Earth's upper atmosphere.

radio source   A source of extraterrestrial radio radiation. The strongest known is Cas A, followed by Cyg A and the Crab Nebula (Tau A) (the capital letters following the name of a constellation refer to the radio sources of the constellation, A being the strongest source). Radio sources are divided into two main categories: Class I, those associated with our Galaxy (which is a weak radio source), and Class II, extragalactic sources. Most radio sources are galaxies, supernova remnants, or H II regions.

radio source counts   The integral number of radio sources per unit solid angle whose measured flux density at the operating frequency of a radio telescope exceeds a certain given value; plot of log N (number of sources) versus log S (where S is in flux units).

radio stars   Stars with detectable emission at radio wavelengths. They include pulsars, flare stars, some infrared stars, and some X-ray stars.

radio window   The wavelength range between a few millimeters and about 20 meters within which Earth's atmosphere is transparent to radiation.

Raman effect   The change of wavelength on scattering. It arises from radiation exciting (or de-exciting) atoms or molecules from their initial states.

Ramsauer effect   An anomalously large mean free path for low-energy electrons.

random walk   If a point experiences successive displacements such that each displacement is in a random direction and of a length also governed by a frequency distribution, then the point is said to experience a random walk. It is a law of statistical behavior closely allied to Brownian motion and the diffusion of molecules. It can be proved that the root mean square displacement experienced in N mean free paths is related to the diffusion coefficient by $D_{rms} = \sqrt{N}$.

Rankine scale   A temperature scale with the same division as the Fahrenheit scale and the zero point at 0° absolute. 0° R = −470° F.

rare gases   The inert gases He, Ne, Ar, etc.

raster   The area of an oscilloscope upon which the image is produced.

rayleigh   Unit of flux. 1 rayleigh = 10^6 photons emitted in all directions per cm² vertical column per second. It is used in measuring the luminous intensity of the aurora.

Rayleigh-Jeans law   An approximation of Planck's blackbody formula (q.v.) valid at long wavelengths ($h\nu \ll kT$). It is often used in radio astronomy; it gives the brightness temperature of a radio telescope.

Rayleigh limit   The minimum resolvable angle (q.v.) between the wavelengths of two spectral lines.

Rayleigh number   A nondimensional parameter involving the coefficients of thermal conductivity and kinematic viscosity which determines when a fluid, under specified geometrical conditions, will become convectively unstable.

Rayleigh scattering   Selective scattering (i.e., preferential scattering of shorter wavelengths) of light by very small particles suspended in the Earth's atmosphere, or by molecules of the air itself. The scattering is inversely proportional to the fourth power of the wavelength (cf. Mie scattering).

Rayleigh-Taylor instability   A type of hydrodynamic instability for static fluids (see Taylor instability) in which the density increases outward.

Razin effect (also called Razin-Tsytovitch effect)   The strong suppression of low-frequency (synchrotron) radiation by electrons moving in a cool, collisionless plasma. It is a theoretical calculation of the Tsytovitch effect (q.v.) specifically directed toward radio astronomy.

recombination   The capture of an electron by a positive ion. It is the inverse process to ionization.

recombination epoch   See matter era.

recombination radiation   See radiative capture.

reconnection   The rejoining of magnetic lines of force severed by the annihilation of the field across the neutral region.

reddening   See extinction.

red giant   A late-type (K or M) high-luminosity (brighter than $M_V = 0$) star that occupies the upper right portion of the H-R diagram. Red giants are post–main-sequence stars that have exhausted the nuclear fuel in their cores. The red-giant phase corresponds to the establishment of a deep convective envelope. Red giants in a globular cluster are about 3 times more luminous than RR Lyrae stars in the same cluster.

red-giant tip   The upper tip of the red-giant branch in the H-R diagram. The red-giant tip represents the "flash" point (e.g., helium flash, carbon flash) where the density and temperature of the core have become high enough that the "ash" in the core is ignited and serves as the fuel for a new series of nuclear reactions. Here a 1 $M \odot$ star ejects its envelope.

redshift (z)   The shift of spectral lines toward longer wavelengths, either because of a Doppler effect (q.v.) or because of the Einstein effect (gravitational redshift). The redshift $z = \Delta\lambda/\lambda$, where λ is the laboratory wavelength of the spectral line and $\Delta\lambda$ is the difference between the laboratory and the observed wavelengths. The redshift of distant galaxies was first noted by Slipher in 1926.

redshift-distance relation   See Hubble's law.

Red Spot   An elliptical spot about 40,000 × 15,000 km on the southern hemisphere of Jupiter. Its color and intensity vary with time. It has been observed for at least a century, and an examination of earlier records shows that Cassini had sketched it in the seventeenth century. Nowadays it is usually interpreted in terms of a Taylor column (q.v.).

reduced proper motion   The observed proper motion of a star (in seconds of arc per year) reduced to absolute proper motion (in kilometers per second).

reflection nebula   A cloud of interstellar gas and dust whose spectrum contains absorption lines characteristic of the spectrum of nearby illuminating stars. The emission component of its spectrum is due to gas; the reflection component, to dust (see also diffuse nebula).

refraction index   See index of refraction.

refraction, law of   See Snell's law.

Regge-Wheeler equations   Schrödinger-type equations for small, odd-parity perturbations on the Schwarzschild metric.

regolith   The layer of fragmentary debris produced by meteoritic impact on the surface of the Moon or a planet.

regression of the nodes   The slow (19.35 per year, 360° in 18.6 years), westward motion of the nodes of the Moon's orbit due to perturbations of the Earth and Sun.

Regulus (α Leo)   A visual triple B8 V star about 26 pc distant.

relative number   See Wolf number.

relativistic (of motions of particles)   Approaching the speed of light.

relativistic bremsstrahlung (gravitational)   Hypothetical gravitational radiation emitted when two stars fly past each other with high velocity and deflect each other slightly.

relativistic zone (of a pulsar)   The region in which M[grams]$/R$[cm] is not negligible compared with unity.

relativity   The special theory concerns time and distance measurements by two observers in uniform relative motion, and clarifies the notion of simultaneity relative to such observers. The general theory of relativity is concerned with the generalization of Newton's law of gravitation when masses moving under their mutual influence acquire velocities comparable to that of light; its basic postulate, derived from the equality of the inertial and the gravitational mass, is that all accelerations are metrical in origin.

relaxation time   Period required for the reestablishment of thermal equilibrium; in particular (in the astronomical context) the period required for the reestablishment of a random distribution of motion in a cluster of stars.

renormalization   The process of proving that a theory gives physically meaningful predictions in all possible applications without violating any basic physical laws.

reseau   A grid that is photographed by a separate exposure on the same plate with star (or galaxy) images.

residual intensity   Ratio of correlated flux in the line to correlated flux in the continuum.

resolving power   The ratio of the mean wavelength of two lines to the minimum resolvable angle (q.v.). The resolving power of the human eye is about 1 minute of arc (its integration time is about $1/15$ second).

resonance   The selective response of any oscillating system to an external stimulus of the same frequency as the natural frequency of the system. Under such conditions the nodes of the two wave trains coincide, and the waves of the initial system increase in amplitude. The natural mode of oscillation of a star varies inversely as the square root of its mean density; if the Sun were to start resonating at its natural frequency, it would have a period of about an hour.

resonance capture   Capture by an atomic nucleus of a particle whose energy is equal to one of the energy levels of the nucleus. Under such circumstances the particle's chances of being captured are greatly increased.

resonance line   The longest-wavelength line arising from the ground state.

resonances   Strongly interacting particles which are born and decay within the short time span of the strong interaction (10^{-23} seconds). The existence of a resonance cannot be observed directly; it can only be inferred from studying the longer-lived products of its decay. An asterisk is commonly used to designate a resonance, e.g., Λ^*.

resonant reaction   A nuclear reaction that has an energetically favorable probability of occurring (see resonance capture).

rest mass   The mass of a body at rest derived consistently with Newton's second law of motion; it is measured in a Lorentz frame in which it is at rest. Photons and neutrinos have a rest mass of zero, but of course they are never at rest and travel at the speed of light for as long as they exist.

restoration   A process used by radio astronomers to eliminate the smoothing effect observed in radio maps that is caused by the finite width of the telescope beam.

reticle   A system of cross-hairs in the eyepiece of a telescope.

retrograde motion   Motion opposite to the prevailing direction of motion.

Reynolds number   A dimensionless number ($R = Lv/\nu$, where L is a typical dimension of the system, v is a measure of the velocities that prevail, and ν is the kinematic viscosity) that governs the conditions for the occurrence of turbulence in fluids.

Rhea   Sixth satellite of Saturn, discovered by Cassini in 1672. Diameter about 1500 km; rotation period $4^d12^h25^m$. Albedo 0.57.

Rigel (β Ori)   A B8 Ia supergiant at least 400 pc distant, the brightest star in Orion. It is a multiple star.

right ascension (R.A. or α)   Angular distance (in hours, minutes, and seconds) along the celestial equator eastward from the vernal equinox to the hour circle of the object. It is analogous to terrestrial longitude.

ring galaxy   See galaxies (de Vaucouleurs classification). According to Freeman and de Vaucouleurs, a ring galaxy results from the collision of a normal spiral with intergalactic gas clouds.

"Ring" Nebula   A planetary nebula (M57, NGC 6720) in Lyra.

rise time (in rocket or balloon astronomy)   The time required for the vehicle to achieve its optimum height.

Ritchey-Chrétien telescope   A system of two mirrors, aspherized to give an image at the secondary (Cassegrain) focus free from spherical aberration and coma.

Ritz combination principle   A principle discovered empirically before the advent of quantum mechanics which states that every spectral line of a given atom corresponds to the difference of some pair of energy levels.

Robertson-Walker metric   A metric which is appropriate for a space-time which is homogeneous and isotropic. It was derived by Robertson and Walker with a minimum of assumptions; but the metric had been used earlier by Friedmann to derive the cosmological models (without a cosmological constant) that are currently in use.

Rocard scattering   Linearly anisotropic scattering.

Roche limit   The minimum distance at which a satellite under the influence of its own gravitation and that of a central mass about which it is describing a circular Keplerian orbit can be in equilibrium. For a satellite of negligible mass, zero tensile strength, and the same mean density as its primary, in a circular orbit around its primary, this critical distance is 2.44 times the radius of the primary. (For the Moon, whose density is lower than that of Earth, the Roche limit would be 2.9 Earth radii.)

Roche lobe   The first equipotential surface for two massive bodies describing circular orbits around one another which forms a figure eight enclosing the two objects. The Roche lobes are the two lenticular volumes enclosing the two bodies.

root mean square (rms)   The square root of the mean square value of a set of numbers (see also random walk).

Rosette Nebula (M16, NGC 2237–2244)   An H II region in Monoceros, more than 1 kpc distant. It has a fairly high degree of symmetry about a centrally located cluster of about six hot, young O-type stars.

Rossby waves   Cyclonic convection waves in a rotating fluid. Such waves occur in the atmosphere, in the oceans, and in the fluid core of the Earth.

Rosseland mean absorption coefficient   A coefficient of opacity which is a weighted inverse mean of the opacity over all frequencies. It is applied when the optical depth is very large.

Rossiter effect   A rotational distortion of the velocity curves of eclipsing binaries.

rotational transition   A slight change in the energy level of a molecule due to the rotation of its constituent atoms about their center of mass.

runaway stars   Stars of spectral type O or early B with unusually high space velocities. The three best known are AE Aur, 53 Ari, and μ Col, all of which diverge from a comparatively small area in Orion. Runaway stars are thought to be produced when there is a supernova explosion in a close binary system.

Runge-Kutta method   A step-by-step method of numerical integration.

Russell-Saunders coupling   See *LS* coupling.

Russell-Vogt theorem   See Vogt-Russell theorem.

rydberg   A unit of energy ($R = \hbar^3 c/me^4$) equal to 13.5978 eV (the ionization potential of hydrogen).

Rydberg correction   See quantum defect.

Rydberg formula   A formula by which the various lines in a given spectral series are obtained: $\lambda^{-1} = R[(m + k_1)^{-2} - (n + k_2)^{-2}]$, where m is an integer, n is any integer greater than m, k_1 and k_2 are empirical corrections which are different for different series, and R is the Rydberg constant, which has a value of 109,678 if λ is measured in centimeters (see also Balmer formula).

S

S band   A radiofrequency band at a wavelength of 11.1 cm.

s-electron   An orbital electron whose l quantum number is zero.

S-factor   A nuclear cross-section factor measured in keV-barns.

S-matrix (scattering matrix)   A matrix (q.v.) representing the transitions from some initial to some final state in a given interaction. The transitions may involve changes in the number of particles in the system.

s-process (slow neutron capture)   A process in which heavy, stable, neutron-rich nuclei are synthesized from iron-peak elements by successive captures of free neutrons in a weak neutron flux, so there is time for β-decay before another neutron is captured (cf. r-process). This slow but sure process of nucleosynthesis which is assumed to take place in the intershell regions during the red-giant phase of evolution, at densities up to 10^5 g cm^{-3} and temperatures of about 3×10^8 K (neutron densities assumed are 10^{10} cm^{-3}). The s-process slowly builds stable nuclear species up to $A = 208$ (time between captures about 10–100 years). It ends there, because any further capture of neutrons leads immediately to α-decay back to lead or thallium. The most likely source of neutrons for the s-process is linked to thermal instabilities in the helium shell during double shell burning after core He exhaustion. The s-process probably occurs in stars where $M < 9$ $M\odot$.

S stars   Red-giant stars of spectral type S are similar to M stars except that the dominant oxides are those of the metals of the fifth period (Zr, Y, etc.) instead of the third (Ti, Sc, V). They also have strong CN bands and contain spectral lines of lithium and technetium. Pure S stars are those in which ZrO bands are very strong and TiO bands are either absent or only barely detectable. Almost all S stars are LPVs. (S1,0. The number following the comma is an abundance parameter.)

S-state, S-level   The state of an atom in which the orbital angular momentum L (the vector sum of the orbital angular momenta l of the individual electrons) is zero.

s-wave (the s stands for secondary)   A seismic shear wave that moves transversely through Earth. The s-waves cannot penetrate the core of the Earth, being totally reflected by the 2900-km discontinuity.

SC stars   Stars which appear to be intermediate in type between S stars and carbon stars (C/O ratio near unity).

SI units   See International System of Units.

FG Sagittae   A supergiant whose spectral type has changed from B4 Ia in 1955 to A5 Ia in 1967 to F6 Ia in 1972. It ejected a planetary nebula some 6000 years ago. It showed s-process elements in its surface layer in 1972 that did not exist in 1965—an indication of deep mixing.

WZ Sagittae   A recurrent DAe old nova (1913 and 1946) with the shortest known orbital period (about 80 minutes). It is almost certainly a close binary system in which mass is being transferred onto a white-dwarf primary.

Sagittarius A   A radio source (the galactic center) about 12 pc in diameter. (Sgr A West is a thermal source; Sgr A East is a nonthermal source.)

Sagittarius arm   One of the spiral arms of the Milky Way, lying between us and the center of the Galaxy in the direction of Sagittarius. It includes the Scutum arm, the 3-kpc arm (q.v.), and the Norma arm. It is about 1.5 kpc from the Sun and about 8.7 kpc from the galactic center. Density of H I and H II in Sagittarius arm is about 1.2 atoms cm^{-3}.

Sagittarius B2   A massive ($3 \times 10^6 M\odot$), dense (up to 10^8 particles per cm^3) H II region and molecular cloud complex—the richest molecular source in the Galaxy. It is in the galactic plane about

10 kpc distant, near the galactic center.

Saha equations   Equations determining the number of atoms of a given species in various stages of ionization that exist in an assembly of atoms and electrons in thermal equilibrium.

Salpeter process   See 3α process.

Saros   A particular cycle of similar eclipses (lunar or solar) known to the Babylonians, that recur at intervals of 6585 days (about 18 tropical years). The interval contains 223 synodic months (6585.32 days) and 19 ecliptic years (6585.78 days). (It also contains 242 nodical months.) The difference of a fraction of a day causes each eclipse to fall about 120° west of the previous eclipse.

satellite lines (of an OH source)   The lines arising from transitions at 1612 and 1730 MHz.

Saturn   Sixth planet from the Sun. Mean distance from Sun 9.540 AU; $e = 0.056$, $i = 2°29'33''$. Sidereal period 29.458 years; synodic period 378 days. Equatorial diameter 116,340 km. Oblateness 0.1. Mass 5.7×10^{29} g $= 95.2$ M_E; mean density 0.7 g cm^{-3}; surface gravity 11 m s^{-2}; V_{esc} 33.1 km s^{-1}. Rotation period at equator 10^h14^m; at poles 10^h38^m. Obliquity 26°44'. T_{eff} about 160 K. V_{orb} 9.65 km s^{-1}. Albedo 0.50. Atmosphere hydrogen and helium . Ten satellites, all of which are locked in synchronous rotation.

Saturn's rings   A system of four concentric rings, only about 2–4 km thick. The outermost ring is ring A, then comes Cassini's division, then ring B (also called the bright ring), then Lyot's division, then ring C (the crepe ring), then ring D (discovered in 1969). The rings are a swarm of solid particles, probably [1973] jagged rocks about 1 meter to 1 km across, not ice as previously had been assumed, inside the Roche limit. Bobrov (1969) estimates the total mass of the rings to be about 0.01 the lunar mass.

Saturn Nebula   A double-ring planetary, NGC 7009, about 700 pc distant.

scalar tensor theory   See Brans-Dicke theory.

scale height   The height at which a given parameter changes by a factor e. For example, an atmospheric scale height of 100 km means that the value at 100 km is $1/e$ the value at the surface.

scaling   In scattering experiments, the property whereby the likelihood of a reaction depends not so much on the amount of energy transferred to the target as on the ratio between energy transferred and momentum transferred.

scattering   Light absorbed and subsequently reemitted in all directions at about the same frequency.

Schmidt telescope   A type of reflecting telescope (more accurately, a large camera) in which the coma produced by a spherical concave mirror is compensated for by a thin correcting lens placed at the opening of the telescope tube. The Schmidt has a usable field of $0°.6$.

Schrödinger's equation   A quantum-mechanical wave equation describing the nonrelativistic motion of a particle or system of particles under the influence of forces. The solutions to Schrödinger's equation yield the wave function describing the system (particle, atom, molecule). This is the fundamental equation in nonrelativistic quantum mechanics.

Schuster mechanism   A scattering mechanism in the continuum, which under certain conditions can yield emission lines in the spectrum even under the assumption of LTE. It is the modification of the emergent radiation, for a given temperature distribution, by variations in the ratio of pure absorption to scattering opacity.

Schwarzschild black hole   A nonrotating, spherically symmetric black hole derived from Karl Schwarzchild's 1916 exact solution to Einstein's vacuum field equations.

Schwarzschild filling factor   Ratio of the actual density to the limiting value for a system.

Schwarzschild radius   The critical radius, according to the general theory of relativity, at which a massive body becomes a black hole, i.e., at which light is unable to escape to infinity. $R_S = 2GM/c^2$; R_S for Sun, 2.5 km; R_S for Earth, 0.9 cm.

scintillation   Variations in the brightness of starlight (i.e., "twinkling") caused by turbulent strata very high in Earth's atmosphere. Scintillation increases with distance from the turbulent zone (cf. seeing).

scintillation counter   A device used with a photomultiplier tube to detect or count charged particles or gamma rays.

α Sco   See Antares.

β Sco   A system with at least five components which during the 1970s is undergoing a series of occultations by the Moon and by Jupiter. Component A is a spectroscopic binary (B0.5 V, B V). In 1971 component C was occulted by Io.

Sco-Cen association   An association of very young stars about 200 pc distant in the Gould Belt (q.v.). The most luminous member is a B star of $M_V = -4.9$.

Scorpius OB1   An extremely young association of OB stars in Scorpius, about 2 kpc distant.

Scorpius X-1 (3U 1617 – 15)   A compact eclipsing X-ray source about 250–500 pc distant. It is the brightest X-ray source in the sky (besides the Sun) and was discovered in 1962. It has day-to-day variations (period about 0.78 days?) of as much as 1 mag; it also has optical and radio counterparts, but no correlation has been found among the flares observed at the three different wavelengths. It is a thermal X-ray source, probably associated with a rotating collapsed star surrounded by an extensive envelope. Tentative optical identification with the 13th mag blue variable V818 Sco. The spectrum of Sco X-1 is similar to that of an old nova.

Scott effect   A selection effect in the study of the magnitude-redshift relation in cosmology. It was pointed out by Elizabeth Scott in 1957 that at great distances only the most luminous clusters of galaxies would be visible, and this fact would introduce a bias into the data.

Sculptor group of galaxies   A group of about five galaxies near the South Galactic Pole, which includes the giant (dusty) spiral NGC 253.

Sculptor system   A dwarf elliptical galaxy ($M_V = -11.28$ mag, mass about $3 \times 10^6 \, M\odot$), about 85 kpc distant, in the Local Group. Discovered in 1938.

δ Scuti stars (also called dwarf Cepheids or ultrashort-period Cepheids)   A group of pulsating variable stars of spectral class A–F with regular periods of 1–3 hours and with small variations in amplitude. They lie in the lower part of the Cepheid instability strip.

second (s)   A unit of time defined as the duration of 9,192,631,770 periods of the radiation corresponding to the transition between the two hyperfine levels of the ground state of the cesium-133 atom. In 1967 the General Conference of Weights and Measures (CGPM) adopted this as the tentative definition of the second in SI units, replacing the ephemeris second, which remains in the IAU system of astronomical constants.

secondary cosmic rays   Atomic fragments—mainly muons—produced by collisions between primary cosmic rays and the molecules in Earth's atmosphere.

secular change   A continuous, nonperiodic change in one of the attributes of the states of a system. Often, a change in an orbit due to dissipation of energy (cf. canonical change).

secular instability   Instability caused by the dissipation of energy.

secular parallax   A parallax based on solar motion; i.e., the baseline is the distance the Sun moves in a given interval of time with respect to the local standard of rest (4.09 AU per year).

secular stability   The condition in which the equilibrium configuration of a system is stable over long periods of time.

seed nuclei   Nuclei from which other nuclei are synthesized.

seeing   The quality of a telescopic image, which is affected by variation in image position and structure due to the unsteadiness of the Earth's atmosphere (cf. scintillation). Seeing is mainly a lower-atmosphere phenomenon and is independent of the distance from the turbulent zone. In good seeing, stellar images are of small diameter; in poor seeing, the images become blurred.

Selected Areas   262 small (75′ square) regions of the sky in which magnitudes, spectral types, and luminosity classes of stars have been accurately measured and which have served as standards for magnitude systems.

selection rule   A rule whereby changes in quantum numbers can take only certain allowed values: e.g., $\Delta l = \pm 1$ or 0 for dipole transitions.

selective absorption   The reddening of starlight in passing through fine particles of interstellar dust.

self-absorption   Reduction in relative intensity in the central portion of spectral lines resulting from selective absorption by a cooler shell surrounding the hot source.

self-consistent field approach   An approach in which the density distribution and state of motion in a system are determined so as to be self-consistent with the force field (e.g., gravitational or electromagnetic) arising from the system itself.

semiconvection   The partial convective mixing that takes place in a convectively unstable region where stability can be attained by the results of the mixing before the region is completely mixed.

semiforbidden lines   Spectral lines from "semiforbidden" transitions, i.e., those whose transition probabilities are perhaps 1 in 10^6 instead of about 1 in 10^9 for forbidden transitions. One bracket—e.g., C III]—is used to indicate semiforbidden lines.

semiregular variable   A class of giant and supergiant pulsating stars of spectral class M, K, N, R, or S with a periodic (or semiperiodic) light curve of varying amplitude. Betelgeuse is one.

sense   One of two opposite directions describable by the motion of a point, line, or surface.

CV Serpentis   A sometimes-eclipsing binary composed of a Wolf-Rayet star and a B0 star with a period of 29.6 days.

λ **Serpentis**   A G0 V star almost identical to the Sun in its energy distribution.

Seyfert galaxy   One of a small class of galaxies (many of which are spirals) of very high luminosity and very blue continuum radiation with small, intensely bright nuclei whose spectra show strong, broad, high-excitation emission lines probably caused by discrete clouds moving at velocities that are higher than the escape velocity. Seyferts possess many of the properties of QSOs, such as the ultraviolet excess of the continuum, the wide emission lines, and the strong infrared luminosity. The energy sources in their nuclei are unexplained; presumably the energy input can be associated with some process that liberates gravitational binding energy to accelerate relativistic particles. Seyferts comprise about 1 percent of the bright galaxies. The brightest Seyfert known is NGC 1068. Weedman-Khachikian classification: class 1 Seyferts have broad Balmer line wings; class 2 have no obvious Balmer line wings.

Seyfert's Sextet   A compact group of galaxies surrounding NGC 6027. It has both spiral and irregular members.

shear   A stress applied to a body in the plane of one of its faces.

sheath   The boundary layer of charged particles between a plasma and its surrounding material.

shell star   A hot main-sequence star, usually of spectral class B–F, whose spectrum shows bright emission lines presumed to be due to a gaseous ring or shell surrounding the star.

shock wave   A sharp change in the pressure, temperature, and density of a fluid which develops when the velocity of the fluid begins to exceed the velocity of sound.

shot noise   Noise, or fluctuations in the current of a detector, due to the fact that the current is carried not by a smooth fluid, but by a large number of individual electrons (cf. wave noise; correlator).

sidelobe   In radio astronomy, a component of the reception pattern of an antenna away from the main beam, representing a direction in which the antenna is sensitive when it should be insensitive.

sidereal day   The length of time ($23^h56^m4\overset{s}{.}091$) between two successive meridian transits of the vernal equinox (cf. mean solar day). Because of precession the sidereal day is about 0.0084 second shorter than the period of rotation of Earth relative to a fixed direction ($23^h56^m4\overset{s}{.}099$).

sidereal period   The time it takes for a planet or satellite to make one complete circuit of its orbit (360°) relative to the stars. Earth's sidereal period (or sidereal year) is equal to 365.2564 mean solar days.

siderite ("star stone")   An iron (or iron and nickel) meteorite. Siderites comprise about 6 percent of known falls.

siderolite (lit.: "sky stone")   A stony iron meteorite. Siderolites comprise less than 2 percent of known falls.

siemens (S)   The SI unit of electrical conduction (reciprocal ohm).

signal band   The wavelength interval within which a feature (e.g., the 21-cm line) is measured (cf. comparison band).

signal-to-noise ratio   The ratio of the amount of intelligible meaning in a signal to the amount of background noise.

singularity (in spacetime)   A region in spacetime where the known laws of physics break down, and the equations become meaningless; a point where one or more components of the Riemann curvature tensor become infinite. The center of the Schwarzschild metric is a singularity in this sense.

sink   In general, a region where energy is given up, in contrast to a source, where energy is released.

Sirius (α CMa)   Also called Dog Star. An A1 V star 2.7 pc distant—the (apparently) brightest star in the sky. Its companion (Sirius B) is a white dwarf of about 0.96 $M\odot$ but only about 0.03 $R\odot$. Period 49.9 years.

skylight   See night-sky light.

slow nova   A nova whose light curve shows a much more gradual development—i.e., rise time of several days, maximum of several weeks, slower decline, amplitude only about 10 mag.

Small Magellanic Cloud   See Magellanic Clouds.

SMC X-1 (2U 0115$-$73)   An X-ray source in the Small Magellanic Cloud. It is a binary system with a 3.89-day period. Identified with Sanduleak No. 160, a B0 I supergiant (m_v = $+13.6$). Because no radial-velocity variations are apparent in Sk 160, the mass of the X-ray emitter must be small relative to Sk 160 (about 2 $M\odot$ if Sk 160 is 20 $M\odot$), unlike the compact member of Cyg X-1.

Snell's law (also called law of refraction)   For a refracted light beam, the ratio of the sine of the angle of incidence to the sine of the angle of refraction is a constant.

snowplow model   A sunspot model in which the expanding current sheath scoops up material like a snowplow, but discards all the accumulated matter when the magnetic field reverses.

Sobieski   Former name of the southern constellation Scutum.

NaI(Tl) detector   Sodium iodide (thallium) detector (also called

NaI(Tl) scintillator). A gamma-ray or X-ray counter developed in 1950.

solar apex   A point on the celestial sphere lying in the constellation Hercules toward which the Sun and the solar system are moving with respect to the local standard of rest at a rate of about 19.4 km per second (about 4.09 AU per year).

solar burst   See burst.

solar constant ($F\odot$)   Energy received per cm^2 per second by a planet at r astronomical units from the Sun (equal to 1.39×10^6 ergs $cm^{-2} s^{-1}/r^2$).

solar cycle   The 11-year period between maxima (or minima) of solar activity. Every 11 years the magnetic field of the Sun reverses polarity; hence the more basic period may be 22 years.

solar neutrino unit (SNU)   1 SNU = 10^{-36} solar-neutrino captures per second per target atom.

solar parallax   Angle subtended ($8''.79$) by the equatorial radius of the Earth at a distance of 1 AU.

solar phase angle (α) (of a planet)   Angular distance at the planet between the Earth and the Sun.

solar system   The Sun and all objects gravitationally bound to it. The solar system is roughly a sphere with a radius greater than 100,000 AU, with the Sun at the center. The Sun is overwhelmingly the dominant object. Planets, satellites, and all interplanetary material together comprise only about 1/750 of the total mass. Geochemical dating methods show that the solar system chemically isolated itself from the rest of the Galaxy $(4.7 \pm 0.1) \times 10^9$ years ago.

solar velocity   Velocity of the Sun (19.4 km s^{-1} in the direction $l^{II} = 51°$, $b^{II} = 23°$) with respect to the local standard of rest.

solar wind   A radial outflow of energetic charged particles from the solar corona, carrying mass and angular momentum away from the Sun. Mean number density of solar wind (1971), 5 per cm^3; mean velocity at Earth 400 km s^{-1}; mean magnetic field 5×10^{-5} gauss; mean electron temperature 20,000 K; mean ion temperature 10,000 K. The Sun ejects about $10^{-13} M\odot$ per year via the solar wind.

soliton (hydrodynamic)   A finite-amplitude disturbance which is propagated through a fluid without any change of shape. MHD solitons are also known.

"Sombrero" galaxy (M104, NGC 4594)   A type Sa/Sb spiral galaxy in Virgo, seen edge-on.

sonde   A rocket or balloon carrying instruments to probe conditions in the upper atmosphere.

source function   The amount of radiant energy per unit mass per unit solid angle emitted in a specified direction. For the case of LTE, it is equal to the Planck function; for pure, isotropic scattering, it is equal to the mean intensity.

South Atlantic Anomaly   A disturbance in the geomagnetic field (a region of intense charged-particle fluxes) over the south part of the Atlantic Ocean. It was discovered in early OAO flights that when the detector passed over that area, the data it collected were not valid.

space charge wave   An electrostatic wave brought about by oscillations of the charges.

spacelike path   A trajectory along which $U \cdot U > 0$.

space motion (space velocity)   Velocity of a star with respect to the Sun; hypotenuse of the right triangle formed by its radial and tangential velocities (cf. peculiar velocity). Space motion vectors are U (in the direction of the galactic anticenter), V (in the direction of galactic rotation), and W (in the direction of the galactic north pole).

spark spectra   The spectra of ions often produced by a spark discharge (cf. arc spectra).

specific gravity   Ratio of the mass of a given volume of a substance to that of an equal volume of water.

specific heat   Ratio of the quantity of heat required to raise the temperature of a unit mass of a substance by a unit amount to that required to raise the temperature of an equal mass of water by the same amount.

specific intensity   See intensity.

spectral class   See Henry Draper system.

spectral energy distribution   The distribution of a star's light among various wavelengths.

spectral index   The power of the frequency to which the intensity at that frequency is proportional. It is positive for thermal radiation, negative for nonthermal radiation.

spectral lines   Discrete emissions (or absorptions) in frequency, usually formed by atomic transitions. The essential difference between optical line spectra and X-ray spectra is that the former correspond to energy changes in the *outer* electrons in an atom, and the latter to energy changes in the *inner* electron orbitals. Gamma rays usually correspond to energy changes in the nucleus. Infrared radiation is produced by high-n transitions of atoms or by the vibration or rotation of molecules. Thermal radio emission is usually produced by still higher-n transitions (the notation 109α corresponds to a transition in a hydrogenic atom between the principal quantum number $n = 109$ and $n' = \Delta n = n+1 = 110$; similarly, a β-transition indicates $\Delta n = 2$, etc.).

spectral series   All spectral lines of a given atom arising from transitions with a common lower energy level.

spectrometer   A spectroscope fitted with a device such as a photoelectric cell for measuring the spectra observed with it.

spectroscopic binaries   Stars whose binary nature can be detected from the periodic Doppler shifts of their spectra, owing to their varying velocities in the line of sight. Double-lined spectroscopic binaries have two sets of spectral features, oscillating with opposite phases. Single-lined spectroscopic binaries have only one set of oscillating spectral lines, owing to the dimness of the secondary component. Spectroscopic binaries are typically of spectral type B, with almost circular orbits (whereas long-period M-type binaries have highly eccentric orbits).

spectroscopic parallax   Parallax for a group of stars based on the magnitudes and spectral types of the member stars. Spectroscopic parallax is by far the most common method of determining stellar distances.

spectrum   Electromagnetic radiation arranged in accordance with its wavelengths in an orderly manner. The spectrum for a given material is a pattern characteristic of its atomic forces, atomic masses, and atomic structure (see spectral lines).

spectrum variables (α^2 CVn stars)   Main-sequence Am or Ap stars whose spectra show anomalously strong lines of metals and rare earths which vary in intensity by about 0.1 mag over periods of about 1–25 days. They are characterized by large magnetic fields (10^3–10^4 gauss) at the surface, small variations in light and color, and small projected rotational velocities. These peculiarities are sometimes interpreted in germs of an oblique rotator (q.v.).

speed-of-light circle   See velocity -of-light radius.

Spica (α Vir)   A double-lined spectroscopic binary (B1.5 V, late B) (period 4.01452 days, $e = 0.146$, $i = 65°$) about 80 pc distant. Component A (10.9 $M\odot$) is a β Cephei star which seems to be near core hydrogen exhaustion.

spicule   A short-lived (about 5 minutes), narrow jet of gas spouting out of the solar chromosphere. Spicules tend to cluster at the edges of supergranulation cells.

spin   The angular momentum of an object with respect to its center of mass; the speed at which it rotates about its own axis.

spin-flip collisions   Collisions between particles in which the direction of the spin angular momentum changes. Since the total angular momentum is conserved, the orbital angular momentum must be changed in magnitude or direction or both. (See 21-cm radiation.)

spinor   A vectorlike quantity with complex components that is used in relativity theory.

spin-up   A discontinuous increase in the pulse frequency of a pulsar.

spiral galaxy   A lens-shaped galaxy with luminous spiral arms of gas, dust, and young stars that wind out from its nucleus. Mass range 10^{10}–10^{12} $M\odot$.

Spitzer-Oort hypothesis   A hypothesis which explains the mass motion of the interstellar gas in terms of the gas pressure gradients existing between H I and H II regions.

Spörer's law of zones   The equatorward drift of average sunspot latitudes.

standard deviation (σ)   The root mean square deviation from the arithmetic mean.

standard error (s.e.)   The standard deviation of a distribution of means or any other statistical measure computed from samples. It is equal to 1.4826 times the probable error.

standing wave   A pattern of oscillations in space in which the regions of maximum displacement and of zero displacement (the nodes) remain fixed in position.

standstill   An interval in the cycle of a variable star during which the brightness temporarily stops fluctuating.

star cluster   A gravitationally bound group containing from several hundred to several hundred thousand stars (see open cluster, globular cluster; see also association).

star streaming   A phenomenon that arises because the mean random speeds of the stars are different in different directions. The direction of star streaming is the direction along which the mean random speed has a maximum value. The phenomenon is caused by the rotation of the Galaxy.

Stark effect   Broadening or splitting of a spectral line caused when a radiating atom or ion is influenced by an electric field, which slightly changes the energy level of the atom. Stark broadening is proportional to the ion and electron density in a plasma and is a good indicator of atmospheric pressure in a stellar atmosphere and hence of the star's luminosity.

stat-coulomb   The unit of charge in the cgs electrostatic system. 1 stat-coulomb $= 3.3 \times 10^{-10}$ coulombs.

static limit   See stationary limit.

static universe   A universe whose radius of curvature is constant and independent of time, as in the Einstein universe.

stationary limit (also called static limit)   In the Kerr solution to Einstein's equations, a surface on which a particle would have to travel at the local light velocity in order to appear stationary to an observer at infinity, and just inside which no particle can remain stationary as viewed from infinity. The stationary limit lies outside the event horizon (q.v.), touching it only at the poles. (In the Schwarzschild solution, the stationary limit coincides with the event horizon.)

stationary wave   A standing wave; the pattern formed when two waves of the same amplitude and frequency move simultaneously through a medium in opposite directions.

statistical equilibrium of a gas   A state in which the average density of atoms per cubic centimeter in any atomic state does not change with time and in which, statistically, energy is equally divided among all degrees of freedom if classical concepts prevail.

statistical error   The uncertainty resulting from a measurement of purely random events. Such an uncertainty is defined as bracketing a range of values within which the correct value has a 66% chance of lying. For example, a value of (100 ± 10) obtained from a given measurement means that the true value has a 66% chance of lying between 90 and 110, and a 34% chance of being either above or below this range.

statistical mechanics   The branch of physics that studies the laws governing systems containing large numbers of particles.

statistical parallax   The mean parallax for a group of stars which are all at approximately the same distance, determined from their radial velocities and from the tau components of their proper motion.

statistical weight (g) (of a state)   The probability that the state will appear under a given set of conditions. Usually, the number of ordinarily degenerate substates contained in the state; e.g., the $(2l+1)m$ states of an atom in the absence of a magnetic field.

steady-state theory   A cosmological theory propounded by Bondi, Gold, and Hoyle in which the Universe has no beginning and no end and maintains the same mean density, in the face of its observed expansion, by the continuous creation of matter at the current rate of 2.8×10^{-46} g cm^{-3} s^{-1} (or roughly one nucleon per cubic kilometer per year). Discovery of the microwave background has persuaded most astronomers to reject the steady-state theory.

Stebbins-Whitford system   See *UBV* system.

Stefan-Boltzmann constant (σ)   The constant of proportionality relating the luminosity of a star to its absolute temperature: $\sigma = 5.67 \times 10^{-5}$ ergs cm^{-2} (deg-K)$^{-4}$ s^{-1}.

Stefan's law   The flux of radiation from a blackbody is proportional to the fourth power of its absolute temperature: $L = 4\pi R^2 \sigma T^4$.

stellarator   A type of plasma machine. It has a twisted-field configuration in the form of a figure 8 to fold the plasma back on itself; therefore, unlike a pinch machine, it has no ends where the plasma can leak out. Stellarators and tokomaks resemble each other in that both are toroidal devices that attain equilibrium and MHD stability through rotational transform and shear; they differ mainly in the way they attain these properties.

step method   See Argelander method.

Stephan's Quintet   A highly disturbed cluster of five peculiar galaxies (NGC 7317, 7318A, 7318B, 7319, 7320) in Pegasus which seem to exhibit gaseous connecting bridges. Four have large redshifts (of the order of 5700–6700 km s^{-1}), but the fifth member (NGC 7320) has a much smaller redshift (800 km s^{-1}). Discovered in 1877 by M. E. Stephan.

sticking coefficient   The fraction of all atoms (e.g., hydrogen) incident on an interstellar dust grain that become adsorbed.

stilb (sb)   1 stilb = 1 candela per square centimeter.

stimulated emission (also called induced emission)   See maser; see also transitions.

stokes (St)   1 St = 1 cm^2 s^{-1}.

Stokes parameters   Four parameters (I, Q, U, V) which must be evaluated to describe fully a beam of polarized light. They involve the maximum and minimum intensity, the ellipticity, and the direction of polarization. The nonvanishing of V indicates the presence of elliptical polarization.

strange particles   The collective name for a group of strongly interacting particles possessing the property of strangeness (q.v.). According to one theory, the strange particles are regarded as the higher quantum states of the nucleus.

strangeness   A property ascribed to certain hyperons (q.v.) whose lifetimes before decay are abnormally long (about 10^{-8} to 10^{-10} seconds) relative to their rates of production (about one every 10^{-23} seconds). Like parity, strangeness is conserved in strong interactions but not in weak ones.

stratosphere   The region of Earth's atmosphere immediately above the troposphere. It starts at a height of about 15 km and goes to a height of about 50 km. The temperature increases from about 240 K to about 270 K.

streamline   Path followed by a moving particle in a fluid when the flow is laminar—i.e., nonturbulent. It is a line in a fluid such that the tangent to it at every point is in the direction of the velocity flow.

Strömgren four-color index (m_1, c_1)   See *uvby* system.

Strömgren sphere   A more or less spherical H II region surrounding a hot star.

strong equivalence principle   In a freely falling and nonrotating laboratory the laws of physics, *including* their numerical content, are the same everywhere including gravity-free space.

strong interaction   The short-range nuclear force which is assumed to be responsible for binding the nucleus together (see interactions). Strong interactions are so called because they occur in the extremely short time of about 10^{-23} seconds. Strong interactions can occur only when the particles involved are less than 3 fermis apart.

SU(3) (symmetrical unitary of order 3)   A symmetry found in subnuclear spectra. It is a concept in group theory, by which Gell-Mann and others, using eight quantum numbers, have been able to combine particles into family groups or supermultiplets, as the lowest-lying eightfold group of the nucleon doublet, the Λ singlet, the Σ triplet, and the Ξ doublet. The SU(3) theory applies only to the strongly interacting particles.

subdwarf (sd)   A star whose luminosity is 1.5 to 2 magnitudes lower than that of main-sequence stars of the same spectral type. Subdwarfs are primarily Population II and lie just below the main sequence on the H-R diagram.

subgiant   A star whose position on the H-R diagram is intermediate between that of main-sequence stars and normal giants of the same spectral type.

subluminous stars   Stars fainter than those on the main sequence. Subluminous stars are stars whose age divided by their life span is close to unity.

subpulse   The weaker component of the pulse of a pulsar.

sum rule   See *f*-sum rule.

Sun   Central body of solar system. Spectral type G2 V. Mass 1.989×10^{33} g; luminosity 3.83×10^{33} ergs s^{-1} of which 2×10^{24} ergs s^{-1} fall on Earth. Radius 695,990 km. Mean density 1.409 g cm^{-3}. Density at surface 3×10^{-7} g cm^{-3}. Rotational period at equator 24^d6^h; at poles, about 35 days. Mean rotation speed 1.9 km

s^{-1}. V_{esc} 618 km s^{-1}; surface gravity 27,398 cm s^{-2}. Surface temperature about 5785 K. Inclination of rotational axis to pole of ecliptic about 7°15′. Central density (Bahcall 1973) 155 g cm^{-3}; central temperature about 14–15 × 10^6 K (both this density and temperature would be lower in a solar model producing a counting rate of less than 1 SNU). Energy generating mass about 0.35 $M\odot$. Galactic orbital period about 220 million years ($e \approx 0$); V_{orb} about 250–300 km s^{-1}. Motion with respect to nearby stars 20 km s^{-1} toward R.A. 18^h4^m, declination $+30°$ (in Hercules). It is about 10 kpc from the galactic center and about 10–15 pc above the galactic plane. $M_v = +4.85$; $M_{bol} = +4.67$. It takes about 1–10 million years for photons to diffuse from the Sun's interior to its surface. About 3% of the energy radiated is in the form of neutrinos. Every second about 655 million tons of H are being converted into 650 million tons of He. A grazing light ray is deflected 1″.7 by the Sun. Magnetic fields about 1–2 gauss over most of its surface; as high as 10–1000 gauss in active regions. If the total angular momentum of the solar system were concentrated in the Sun, its equatorial rotation speed would be about 100 km s^{-1}.

sunspot   A temporary disturbed area in the solar photosphere that appears dark because it is cooler than the surrounding areas. Sunspots usually occur in pairs of opposite polarity about 30° N and S of the equator, and move in unison across the face of the Sun as it rotates. The leading (or preceding) spot is called the p-spot; the following, the f-spot. Some sunspots have magnetic felds as high as 1000 gauss (highest observed was 5000 gauss [Steshenko 1967]). Typical diameter, 10^9 cm.

sunspot number   See Wolf number.

sunspot radiation   Intense, variable, circularly polarized radio waves in a noise storm.

supercluster   A cluster of clusters of galaxies. Scale is about 50 Mpc.

supergiant   An extremely luminous star of large diameter and low density. No supergiants are near enough to establish a trigonometric parallax.

supergranulation cells   Convective cells (about 15,000–30,000 km in diameter) in the solar photosphere, distributed fairly uniformly over the solar disk, that last as long as a day. New sunspots develop in the intersections of adjacent supergranulation cells. Most of the magnetic flux through the photosphere is concentrated in the supergranule boundaries.

super-metal-rich   Richer in metals than the Hyades.

supermultiplet   A multiplet of multiplets.

supernova   A gigantic stellar explosion in which the star's luminosity suddenly increases by as much as a billion times. Most of the star's substance is blown off, leaving behind, at least in some cases, an extremely dense core which (as in the Crab Nebula [q.v.]) may be a neutron star. Supernovae are of two main types: Type I ($M_V = -14$ to -17) have a nonhydrogen spectrum, lower mass, and high velocity (about 10,000 km s^{-1}), and may be produced by the thermonuclear detonation of a highly degenerate core. Type I supernovae are found in both spiral and elliptical galaxies. Type II ($M_V = -12$ to -13.5) have a hydrogen spectrum, higher mass, and lower velocity (about 5,000 km s^{-1}), and occur in young, massive stars near the edge of spiral arms. Type II supernovae are more common: Tammann (1974) finds that Type II supernovae occur in our Galaxy at the rate of 0.01 to 0.05 per year. (Type III supernovae are similar to Type II but are probably of much higher mass.) Novae release about 10^{44} ergs of energy; supernovae, about 10^{49} to 10^{51} ergs.

supernova remnant (SNR)   A gaseous nebula (q.v.), the expanding shell ejected by a supernova, and deriving its energy (at least in some cases) from the conversion by the remanent neutron star of its rotational energy into a stream of high-energy particles being continually accelerated in the SNR. About 100 SNRs are known in our Galaxy. Supernova remnants are usually powerful radio sources.

suprathermal   High-energy.

suprathermal proton bremsstrahlung   Ordinary electron-proton bremsstrahlung viewed from the rest frame of the electron rather than the proton; in other words, the electron is at rest and the heavy particle (proton) is moving.

surface gravity (g)   Also called acceleration due to gravity. The rate at which a small object in free fall near the surface of a body is accelerated by the gravitational force of the body, $g = GM/R^2$. Surface gravity of Earth is equal to 980 cm s$^{-2} \approx 32$ feet s^{-2}.

Swan bands   Spectral bands of the carbon radical C$_2$ first investigated in 1856 by W. Swan. They are a characteristic of carbon stars. Swan bands pass through a minimum between spectral types R4 and R6 and increase again toward N6.

Swan Nebula   See Omega Nebula.

symbiotic stars   A term originally used by P. Merrill to describe stars of two essentially dissimilar kinds which seem to occur together and which seem to "need" each other. In practice, it has come to signify a peculiar group of objects (usually spectral type Me) that display a combination of low-temperature absorption spectra and high-temperature emission lines. These objects undergo semiperiodic nova-like outbursts and display the spectral changes of a slow nova superposed on the features of a late-type star. Their spectra are midway between those of planetary nebulae and true stellar objects. A symbiotic star is now usually taken to be a small, hot, blue star surrounded by an extensive variable envelope. As of 1973 about 30 were known.

synchronous rotation   Rotation whose period is equal to the orbital period.

synchrotron radiation (sometimes called magnetic bremsstrahlung)   The radiation emitted by ultrarelativistic charged particles that are circulating in strong magnetic fields. The acceleration of the particles causes them to emit radiation. A characteristic of such radiation is that it is polarized, and the wavelength region in which the emission occurs depends on the energy of the electron—e.g., 1 MeV electrons would radiate mostly in the radio region, but GeV electrons would radiate mostly in the optical region.

synodic month   The period of time (29.53 days) between two successive identical phases of the Moon—e.g., new Moon to new Moon or full Moon to full Moon (cf. lunation).

synodic period   The period of revolution of one body about another with respect to the Earth. (synodic period)$^{-1}$ = ± (sidereal period)$^{-1}$ ∓ (Earth's period)$^{-1}$.

System I and System II longitude (Jupiter)   In the case of Jupiter, because of its differential rotation, two different rotation states are used to keep track of the cloud markings: $9^h50^m30^s$ for the equator (System I) and $9^h55^m41^s$ for the high latitudes (System II). Since many of the apparently localized sources of radio noise on Jupiter near a wavelength of 15 m have a shorter period than System II for optical nonequatorial features, the IAU has officially adopted a System III ($9^h55^m29^s$) for radio astronomy.

system noise   The noise in a radio telescope, composed of the receiver noise and the sky noise.

T

T associations   Associations (q.v.) containing many T Tauri stars. About 20 are known.

t-time   A time scale in which the relative motion of two observers is nonzero but unaccelerated (see τ-time).

tachyons   Hypothetical faster-than-light particles with imaginary mass, real energy, real momentum, and spacelike.

tail (of a comet)   The long streamer (about 10^7 km long; density about 10^{-18} atm) behind the comet head. Type I tails are straight (ionic tails); type II tails are curved (dust tails, little or no charge). Dust tails are usually driven by radiation pressure; ionic (gas) tails are driven by the solar wind. Comet tails do not usually appear until the comet is inside the orbit of Mars.

tangential velocity   That component of a star's velocity (with respect to the Sun) that lies at right angles to the line of sight (in km s^{-1}; cf. proper motion).

Tarantula Nebula   See 30 Doradus Nebula.

τ-time   A time scale in which there is no relative motion between two observers (cf. *t*-time).

α Tauri   See Aldebaran.

NML Tau = IK Tau   An infrared source (an M-type Mira variable with a period of 465 days) discovered by Neugebauer, Martz, and Leighton in 1965.

RV Tauri stars   A class of about 100 semiregular variable yellow supergiants of late spectral type (G–K), similar to W Virginis stars but with longer periods. Their spectra often contain emission lines, and their light curves have alternating deep and shallow minima. They have a large infrared flux. RV*a* stars maintain an approximately constant mean brightness; RV*b* stars have long-term (on the order of 1000 days) periodicity.

T Tauri stars (sometimes called RW Aurigae stars)   Eruptive variable subgiant stars associated with interstellar matter and believed to be still in the process of gravitational contraction on their way to the main sequence. They are found only in nebulae or very young clusters. They have low-temperature (G–M) spectra with strong emission lines and broad absorption lines. Their absolute magnitudes are brighter than those of main-sequence stars of similar spectral types. They have a high lithium abundance. T Tau itself is dG5e.

Taurus A source   See Crab Nebula.

Taylor column   A column that occurs over a fixed region in a rotating fluid because of the two-dimensional character of the motion in the absence of viscosity. This phenomenon is used to interpret the Red Spot of Jupiter.

Taylor instability   A hydrodynamic instability which occurs whenever there is a density inversion. This configuration is said to be Taylor unstable (or Rayleigh-Taylor unstable) against perturbations that would cause mixing of layers of unequal densities.

technetium   An unstable element which does not exist naturally on Earth. The longest-lived isotope ^{97}Tc has a half-life of 2.6 million years. (Only ^{99}Tc, half-life 2.1×10^5 years, can be produced by the s-process.) Technetium is found only in the spectra of MS, S, and N variable stars.

tektite   A small glassy body containing no crystals, probably of meteoritic origin and bearing no antecendent relation to the geological formation in which it is found.

telluric lines   Spectral lines or bands that originate from absorption by gases such as O_2, H_2O, or CO_2 in the Earth's atmosphere.

temperature   A measure of the average kinetic energy of the particles of a system.

tempon   A unit of time equal to the length of time it takes light to cross the classical radius of an electron (about 10^{-23} seconds).

tera-   A prefix meaning 10^{12}.

terminator   The line of sunrise or sunset on the Moon or a planet (cf. limb).

tesla (T)   The derived SI unit of magnetic flux density. $1 \text{ T} = 1 \text{ Wb m}^{-2} = 10^4$ gauss.

Tethys   Fourth satellite of Saturn, discovered by Cassini in 1684. Diameter about 1000 km; $P = 1.87$ days.

Themis   A satellite of Saturn discovered by Pickering in 1900, but since lost.

thermal bremsstrahlung   A mode of X-ray production by very energetic electrons accelerated in the field of a positive ion.

thermal energy   Energy associated with the motions of the molecules, atoms, or ions in a substance.

thermal equilibrium   Strictly, that equilibrium which is attained by a system that is in contact with a thermal bath at some constant temperature. In such an equilibrium the velocity distributions, for example, are described by Maxwell's law.

thermal equilibrium, law of   The temperature of a body in equilibrium is the same at all points (also called zeroth law of thermodynamics).

thermal noise   See Johnson noise.

thermal radiation   Blackbody radiation; radiation caused by the high temperature of the radiating objects, as opposed to nonthermal radiation, which is caused by energetic (not necessarily hot) electrons.

thermalization   An atomic or molecular transition is thermalized when the Boltzmann factor for the two levels of the transition takes on the value it would have in thermodynamic equilibrium.

thermion   An ion, either positive or negative, which has been emitted by a heated body. Negative thermions are electrons.

thermodynamic equilibrium   The condition of a system whose members have conformed to the principle of equipartition of energy, so that there is no net exchange of energy.

thermodynamics, laws of   The first is the law of conservation of energy; the second is the law of entropy. (See also Nernst theorem; thermal equilibrium, law of.)

thermohaline convection   A type of hydrodynamic instability.

theta pinch   A fusion device in which the magnetic field runs parallel to the plasma column. It is a long cylindrical tube enclosed in a one-turn magnet coil.

thin-screen model   A model in which Gaussian angular scattering is concentrated near one point along the path.

Thirring effect   An effect predicted by general relativity, which causes the dragging of the inertial frame outside a rotating mass. As a pulsar, for example, rotates, it drags along the inertial frames, both inside and outside. (See also Lense-Thirring effect.)

Thomas-Fermi theory   A theory of the energy of partially ionized matter in the limit of high density (cf. Boltzmann-Saha theory).

Thomson scattering   The limit of Compton scattering at low energies.

3α process (or triple-α process)   A nuclear reaction ($3 \,^4\text{He} \rightarrow \,^{12}\text{C} + \gamma + \sim 7$ MeV) by which helium is transformed into carbon. The process is dominant in red giants. At a temperature of about 2×10^8 K and a density of 10^5 g cm^{-3}, after core hydrogen is exhausted, three α-particles can fuse to form an excited nucleus of ^{12}C, which occasionally decays into a stable ^{12}C nucleus. The overall process can be looked upon as an equilibrium between three ^4He nuclei and the excited $^{12}\text{C}^*$, with occasional irreversible leakage out of the equilibrium into the ground state of ^{12}C. Further capture of α-particles by ^{12}C nuclei produces ^{16}O and ^{20}Ne. (The details of this process were worked out by Salpeter and Hoyle.)

3-kpc arm   A component of the Sagittarius arm (q.v.) with noncircular gas motions. It is seen in absorption against Sgr A with a velocity of -53 km s^{-1}, implying that at least part of the arm is expanding away from the galactic center. The nearest "edge" is presently at a radius of 4 kpc from the galactic center. (Star formation can be triggered by an outward-moving shock wave.)

threshold energy   Difference between the energy at the first excited level and that of the ground state.

Thuban (α Dra)   A fourth-mag A0 star. It was the "Pole Star" at the time the Egyptians built the Pyramids.

time delay   See dispersion.

timelike path   A path whose tangent obeys $U \cdot U < 0$. In relativity, all material particles travel along timelike paths.

tincle (track in the cleavage)   The track of a charged particle in a meteorite.

Titan (S VI)   The largest and brightest (albedo 0.21) satellite of Saturn, discovered by Huyghens in 1655. $R \approx 2900$ km (about the size of Mercury), period (orbital and spin) $15^d 22^h 41^m$. H_2 and CH_4 have been discovered in its atmosphere.

Titania   Fourth satellite of Uranus, discovered by Herschel in 1787, $R \approx 850$ km; $P = 8^d 17^h$.

tokamak (the name is a Russian acronym)   A type of "magnetic bottle" used in experiments on controlled nuclear fusion.

Toro   An Earth-crossing asteroid (No. 1685) discovered by Wirtanen in 1948 and rediscovered in 1964, of irregular shape (about 5 × 3 km), whose closest approach takes it within 0.13 AU of Earth. Perihelion distance 0.77 AU; aphelion distance 1.96 AU; $e = 0.44$. Orbital period 584.2 days (⅘ that of Earth); rotation period $10^h 11^m$; $a = 1.37$ AU. Radar observations indicate a rocky surface, thinly covered with dust. High albedo (≤ 0.15).

torr   A unit of pressure equal to 1/760 of an atmosphere, or about 1 mm Hg.

transient X-ray sources   As of early 1974, four had been detected: Cen X-2, Cen X-4, 2U 1543−47, and Cep X-4. They resemble slow novae.

transit (of a star)   The passage of a star across the meridian; or the passage of an inferior planet across the Sun's disk.

transit-time effect   The time required for the radiation to travel from the source to the object which reflects, or absorbs and reemits, it to the observer.

transition probability   The probability that a system in one energy state will undergo a transition into another. Associated with any given pair of energy levels are three transition probabilities: the spontaneous-emission probability, the absorption probability, and the stimulated-emission probability.

transition radiation   Radiation emitted (in the X-ray region) when energetic charged particles pass through an interface between two media of different dielectric properties.

transverse velocity   See tangential velocity.

transverse waves   Waves vibrating at right angles to the direction of propagation—e.g., electromagnetic waves.

Trapezium (θ^{1} Ori)   Four very young stars (O6–B3) (there are six stars in the system) in the center of the Orion Nebula. They form the vertices of a trapezoid.

trapped surface   A surface (e.g., of a black hole) from which light cannot escape to infinity.

Triangulum Galaxy (M33, NGC 598)   A Sc II–III spiral galaxy, a satellite of the Andromeda Galaxy, about 700 kpc distant. $M_V = -18.9$.

Trifid Nebula (M20, NGC 6514)   An emission nebula in Sagittarius, ~ 1 kpc distant.

triple-α process   See 3α process.

tritium (T)   A heavy isotope of hydrogen, whose nucleus is composed of one proton and two neutrons. It does not exist naturally on Earth, and its half-life is 12 years. Mass of tritium atom, 3.016 amu.

triton (τ)   The nucleus of the tritium atom (q.v.).

Triton   The inner satellite of Neptune, discovered by Lassell in 1846. It is larger than the Moon (R $\approx$ 2900 km), with an almost circular retrograde orbit of 5 days 21 hours.

Trojans   The asteroids located at the points of Jupiter's orbit around the Sun that are equidistant from the Sun and Jupiter (see Lagrangian points). The first Trojan (Achilles) was discovered in 1906. About 15 are now known.

tropical year   The interval of time between two successive vernal equinoxes. It is equal to 365.242 mean solar days.

tropopause   Upper boundary of the troposphere (about 15 km), where the temperature gradient goes to zero.

troposphere   Lowest level of Earth's atmosphere, from zero altitude to about 15 km above the surface. This is the region where most weather occurs. Its temperature decreases from about 290 K to 240 K.

Trumpler stars   A class of extremely luminous (and formerly considered extremely massive) stars.

Tsytovich effect   An effect wherein the index of refraction of a medium is much less than unity so that the phase velocity of electromagnetic waves is greater than the speed of light in the medium.

In this case, a relativistic electron can no longer keep in phase with the waves it generates, and the intensity of synchrotron radiation is very much reduced. (See also Razin-Tsytovitch effect.)

47 Tucanae   A metal-rich globular cluster about 5.1 kpc distant. It has roughly one-quarter the solar metal abundance. It has a high galactic latitude and low reddening.

tunneling   A phenomenon in quantum mechanics whereby a particle has a nonzero chance of crossing a potential barrier into a region which in classical mechanics would be forbidden to it. Tunneling is a direct consequence of the wave nature of material particles.

turbidity   The appearance of a star as a disk in a long-exposure photograph, due to the scattering of adjacent continuum light by emulsion grains.

turnoff point   The point on the H-R diagram at which a cluster of stars turns off from the main sequence. The higher the turnoff point, the younger the cluster.

21-cm radiation   The emission line (in the radio range at a frequency of 1420 Mc/s) of neutral (atomic) hydrogen, caused when an electron "flips" from spinning in a direction parallel to the proton's spin to the opposite direction of lower energy. Spontaneous transitions from one level to another occur only once every 11,000,000 years on the average for a single electron, but there are so many billions of electrons in the Milky Way that this radiation can be detected by radio telescopes. (The analogous deuterium line is at 91.6 cm.) First detected in 1951; 2 years later extragalactic H I was detected.

two-component model   A model of the solar wind which has two thermal components—electron and proton gases of differing temperatures.

Tycho's star (3C 10)   Remnant of a Type I supernova (B Cas), 3–5 kpc distant, which Tycho observed and described in 1572. At its peak it was as bright as Venus and was visible in the daytime, reaching a magnitude of about -4. It is an X-ray source (2U $0022 + 63$).

U

U line   A sodium line at 3302 Å.

***UBV* system**   A system of stellar magnitudes devised by Johnson and Morgan at Yerkes which consists of measuring an object's apparent magnitude through three color filters: the ultraviolet (U) at 3600 Å; the blue (B) at 4200 Å; and the "visual" (V) in the green-yellow spectral region at 5400 Å. It is defined so that, for A0 stars, $B - V = U - B = 0$; it is negative for hotter stars and positive for cooler stars. The Stebbins-Whitford-Kron six-color system (U, V, B, G, R, I) is defined so that $B + G + R = 0$.

***Uhuru* satellite (Small Astronomy Satellite A)**   A satellite devoted entirely to the study of cosmic X-ray sources. It was launched off the coast of Kenya on 1970 December 12.

ultrarelativistic (of energetic particles)   Having velocities very nearly equal to the velocity of light ($E >> mc^2$).

ultrashort-period Cepheids   See Delta Scuti stars.

ultraviolet-bright stars   Stars that are brighter than the horizontal-branch stars and bluer than the giant-branch stars.

ultraviolet radiation   Electromagnetic radiation "beyond the violet" with wavelengths in the approximate range 100–4000 Å.

ultraviolet stars   Very hot pre–white-dwarf stars; usually the hot central stars of planetary nebulae which are contracting toward the white-dwarf state.

Umbriel   A satellite of Uranus about 400 km in diameter (period 4.1 days). Discovered by Lassell in 1851.

***Umklapp* scattering**   The contribution to scattering caused when the exchange of momentum crosses the boundary of a Brillouin zone.

uncertainty principle   The principle that the fundamental uncertainty in a variable times that in its canonical conjugate is of the order of Planck's constant: $\Delta x \Delta p = h$. Thus the uncertainty in the measurement of the position of an electron varies inversely as the uncertainty in the measurement of its momentum. A corollary is that it is impossible to measure an atomic or nuclear process without at the same time disturbing or altering the process.

unitarity   The principle of conservation of probability. An example would be that if a particle can decay by several modes, the sum of the fractions taking each particular mode should add up to one.

unitary transformation (U)   A transformation whose reciprocal is equal to its Hermitian conjugate.

Universal Time (UT)   The local mean time of the prime meridian. It is the same as Greenwich mean time, counted from 0 hours beginning at Greenwich mean midnight. UT0 is uncorrected; UT1 is corrected for the Chandler wobble; UT2 is corrected both for the Chandler wobble and for seasonal changes in Earth's rotation rate.

Universal Time Coordinated (UTC)   See Coordinated Universal Time.

universe   The total celestial cosmos. According to Gott *et al.* the universe seems to be on a large scale isotropic, homogeneous, matter-dominated, and with negligible pressure. The total proper mass content of about 10^{23} $M \odot$ (Sandage derives 10^{56} g from his determination of the deceleration parameter q_0) and radius of about 2×10^{28} cm are the order of magnitude that most cosmologists would accept if the universe is bounded. Total mass contributed by luminous matter, about 3×10^{53} g (see mass discrepancy). Age about 18×10^9 yr for a Hubble constant $H_0 = 55$ km s^{-1} Mpc^{-1}.

Uranus   Seventh planet from the Sun, discovered by Herschel 1781 March 13. Mass 8.78×10^{28} g; radius 25,400 km; oblateness 0.07. Mean density 1.21 g cm^{-3}. Rotation period $10^h49^m26^s$ retrograde. Mean distance from Sun 19.18 AU. Orbital period 84.0 years; orbital velocity 6.8 km s^{-1}; $e = 0.04$; $i = 0°.8$; obliquity $97°.9$. Escape velocity 22 km s^{-1}; surface gravity 0.96 Earth's. Synodic period 369.66 days. Albedo 0.66. Maximum brightness $+5.7$ mag. Surface temperature about 110 K. Atmosphere H_2 and CH_4. Five satellites, all of which orbit in its equatorial plane.

Urca process   A series of nuclear reactions, primarily among the iron group of elements, accompanied by a high rate of neutrino formation and postulated as a cause of stellar collapse. Neutrinos carry away energy quickly and invisibly, so this process was named for the Urca casino in Rio de Janeiro, which carried away money the same way.

W Ursae Majoris stars   A large class of double-lined eclipsing binaries with very short periods (a few hours) whose spectra indicate mass transfer. They are distinguished by the fact that their primary and secondary minima are equal. They are all F or G binaries on or near the main sequence. They may be the progenitors of dwarf novae.

Ursa Minor system   An intrinsically faint ($M_V \approx -9$) dwarf elliptical galaxy about 70 pc distant, in the Local Group.

UV stars   See ultraviolet stars.

uvby **system**   A four-color intermediate-band stellar-magnitude system devised by Strömgren, consisting of measures in the ultraviolet, violet, blue, and yellow regions.

V

Van Allen belts   Two doughnut-shaped belts in the Earth's magnetosphere (inner belt some 3000 km above the surface; outer belt, 18,000–20,000 km above the surface), where many energetic charged particles from the solar wind are trapped in Earth's magnetic field. The energy of the particles is highest in the inner belt.

van Biesbroeck's star (Gliese 752b, BD + 4°4048B)   A very faint (M_v = 18.6; M_{bol} = 13.12), nearby (parallax $0''.168$, about 8 pc distant) dM5e star of *very* low mass (0.07 $M\odot$). Temperature about 2250 K.

van der Waals forces   The relatively weak attractive forces operative between neutral atoms and molecules.

van Maanen's star   A white dwarf 4 pc distant; density 4×10^5 g cm^{-3}.

variable star   A star that varies in luminosity. The first variable discovered in a given constellation has the letter R preceding the name of the constellation. Then S, . . . , Z. Then RR, RS, . . . , RZ, SS, . . . , SZ, . . . , ZZ. Then AA, . . . , AZ (the letter J is never used), BB, . . . , BZ, . . . , QQ, . . . QZ. The next variable (the 335th) is given the designation V335. (See also Cepheids, flare stars, long-period variables, novae, etc.)

vector meson   See intermediate vector boson.

vector translation   The small theoretical precession of the axis of an orbiting body due to the gravitational influence of its primary. This effect is predicted by general relativity, but so far it has not been observed.

Vega (α Lyr)   An A0 V star about 8 pc distant. It is the standard A0 V star in the *UBV* system.

Veil Nebula   See Cygnus Loop.

Vela pulsar (PSR 0833 − 45)   A pulsar about 400–500 pc distant, probably associated with the Vela supernova remnant. Period 0.0892 seconds.

Vela **satellites**   A sequence of satellites launched to monitor possible violations of the nuclear test ban treaties. The system consists of four satellites in a circular orbit around the Earth with a radius of 120,000 km. The *Vela* satellites have detected cosmic gamma-ray bursts (q.v.).

Vela supernova remnant   A gaseous nebula in the middle of the Gum Nebula, the remnant of a Type II supernova whose light reached Earth about 10,000–30,000 years ago. It consists of bright filaments that form a D-shaped ring in Hα and a rough circle in the ultraviolet. It includes the Vela X, Y, and Z radio complexes and is a strong X-ray source.

Vela X   A compact radio source about 400–500 pc distant associated with the Vela supernova remnant (q.v.). It has a nonthermal radio spectrum and is about 20 percent polarized. It is associated with the Gum Nebula, the Vela pulsar, and the X-ray source 2U 0832−45, although the pulsar and the X-ray source are displaced about $0°.7$ from the center of the Vela X radio emission. Vela Y and Vela Z are outlying components, also nonthermal, but too weak to exhibit polarization.

Vela X-1 (3U 0900−40)   An eclipsing X-ray source identified with the seventh-magnitude single-lined spectroscopic binary HD 77581 (B0.5 Ib) with a period of 8.96 days. Estimated mass of unseen companion 1.7–15 $M\odot$, with a probable value of about 2.6 $M\odot$.

velocity dispersion   Random motion of galaxies in a cluster.

velocity-distance relation   See Hubble's law.

velocity-of-light radius (also called velocity-of-light cylinder)   The radius of a rotating neutron star at which the rotational velocity of the plasma approaches the velocity of light.

velocity profiles   In radio astronomy, the output response for all filters for a given position of the beam on the source (cf. drift curves).

velocity space   The subspace of phase space whose coordinates are the velocities in each of the three directions of ordinary space.

AI Velorum stars   A class of dwarf Cepheids. They are all RR Lyrae stars with periods shorter than 0.25 days.

γ² Velorum   A triple system (WC8, B1 IV, O9 I) embedded in the Gum Nebula, probably about 400 pc distant. Period 78.5 days. It is the brightest Wolf-Rayet star in the sky ($M_V = -5.6$).

Venus   Second planet from the Sun. Mass 4.872×10^{27} g; radius of solid surface 6056 km; radius of cloud surface 6100 km. Mean density 5.16 g cm^{-3}. V_{esc} 10.3 km s^{-1}; surface gravity 8 m s^{-2}. Surface temperature (from *Venera 8*) 743 ± 8 K; temperature of cloud tops about 250 K. Mean distance from Sun 0.7233 AU;

orbital period 224.7 days (synodic period 583.9 days); e = 0.0068, i = 3°39. Rotation period 243.09 ± 0.5 days retrograde (*Mariner 10* has established that the cloud tops rotate every 4 hours retrograde). Obliquity 3° R. Orbital velocity 35 km s^{-1}. Radar experiments have established that the surface is somewhat smoother than the Moon, but there are mountains and there is extensive cratering. Atmospheric pressure 92–95 atm. Atmosphere (by volume 1972) 90–95% CO_2, remainder primarily N_2, traces of water vapor, oxygen, HF, HCl. Maximum elongation 48°. Last transit of Sun was in 1882; next one will be 2004. Venus's rotation period is in synchronism with Earth—that is, at inferior conjunction the same side is always toward the Earth. Albedo 0.76.

vernal equinox   The point of intersection between the ecliptic and the celestial equator, where the Sun crosses from south to north. It is sometimes called the First Point of Aries because several thousand years ago it was in Aries. Because of precession it has now slid west into Pisces and in 200–300 years it will edge into Aquarius. By definition, the vernal equinox is at α = 0°, δ = 0°.

vertex   See radiant.

very large array telescope   A radio telescope scheduled to be built near Socorro, New Mexico, which will consist of 27 antennas, each 82 feet in diameter, distributed along three 13-mile-long arms of a Y-shaped track. According to the NSF, the array will give radio astronomers as much resolution as the 200-inch gives optical astronomers.

very long baseline interferometry (VLBI)   In radio astronomy, a system of two or more antennas placed several hundred or several thousand miles apart, which are operated together as an interferometer.

Vesta   An asteroid 500 km in diameter (P = 1325 days, a = 2.361 AU, e = 0.09, i = 7°1). It is the brightest of all minor planets, at times approaching naked-eye visibility (mag 5.5). Rotation period 5^{h}20^{m}31^{s}665. (Its spectrum can also be interpreted to mean a rotation period of 10^{h}40^{m}58^{s}84.) Albedo 0.24. Discovered by Olbers in 1807.

vibrational energy   Motion of the pair of nuclei in a diatomic molecule along the direction of the internuclear axis (cf. rotational energy).

vibrational transition   A slight change in the energy level of a molecule due to its vibration. If the possibility of the rotation of the molecule as a whole is disregarded, then one gets from each electronic level a sequence of vibrational levels corresponding to various degrees of vibration of the nuclei around their equilibrium position. These are distinguished by the vibrational quantum number v.

vignetting A systematic error in the measurement of stellar magnitudes when the object being measured is far off axis.

violent galaxy (also called explosive galaxy) A type of galaxy differentiated only recently. Violent galaxies include QSOs and exploding galaxies like M82. About 1 percent of the galaxies are classified as violent. Violent galaxies release on the average 10^{58} ergs of energy, compared with a supernova release of 10^{49} ergs. Nearest violent galaxy is Cen A.

α Virginis See Spica.

W Virginis stars See Cepheids.

Virgo A (3C 274) A strong radio source. Optically, it is an elliptical galaxy (M87) with a luminous blue jet about 1500 pc long. It is also an X-ray source (Virgo X-1, 2U 1228 + 12).

Virgo cluster An irregular cluster of about 2500 galaxies ($z = 0.004$), including the giant elliptical M87 (the galaxy of greatest known mass). (Sandage 1974 derives a distance of 19.5 $\pm$ 0.8 Mpc.)

Virgo supercluster A cluster of clusters of galaxies about 30 Mpc in diameter, of which there is evidence that the Local Group forms a part. If so, then the Local Group is eccentrically located about 10 Mpc from the center.

Virgo X-1 (2U 1228 + 12) An X-ray source identical to Virgo A. It is also one of the most powerful extragalactic sources of radiation at infrared wavelengths.

virial theorem For a bound gravitational system the long-term average of the kinetic energy is one-half of the potential energy.

virial-theorem mass The mass of a cluster of stars or galaxies in statistical equilibrium derived by using the virial theorem that the mean square velocity of all the stars or galaxies in a cluster is proportional to the mass of the cluster divided by its radius.

virtual particle A particle that exists for an extremely short time in an intermediate stage of a reaction or transition.

visibility function The Fourier transformation of a distant radio source, normalized to its value at small antenna spacings.

visual binary star See binary system.

visual magnitude The magnitude determined with the eye.

vis viva equation An equation governing the conservation of angular momentum.

Vlaslov equation A collisionless Boltzmann equation, which describes stars moving in regular orbits in an averaged self-contained gravitational field.

Vlasov-Maxwell equations   Equations that describe the propagation of radiation in hot, collisionless plasmas.

Vogt-Russell theorem   If the pressure, the opacity, and the energy generation rate are functions of the local values of density, temperature, and the chemical composition only, then the structure of a star is uniquely determined by the mass and the chemical composition. (When isothermal cores occur in the interiors of stars, then multiple-valued solutions become possible.)

Voigt profile   Profile of a spectral line allowing for the effects of Doppler broadening combined with a Lorentz (damping) profile.

von Zeipel's theorem   The surface brightness of a rotating star or a component of a binary at any point on its surface is proportional to the local value of gravity.

Vulcan   The name of a hypothetical planet at one time thought to exist between the Sun and Mercury.

W

W3   A dense cloud of gas about 3 kpc distant in the Perseus arm.

W boson   See intermediate vector boson.

W44   A radio source. It is a supernova remnant about 3 kpc distant and less than $0\overset{\circ}{.}5$ from the galactic plane.

W49   A radio source (a giant H II region) about 14 kpc distant. It is the most powerful thermal radio source known in our Galaxy.

W51   A radio source, a supernova remnant. PSR $1919+14$ lies within its radio contours.

watt   The SI unit of power. $1 \text{ W} = 10^7 \text{ ergs s}^{-1}$.

wave function (ψ)   A mathematical function that describes the wave-mechanical state of a system (atomic or nuclear). In a one-electron atom, it yields the likelihood that the electron will be found in the neighborhood of that point (per unit volume). This interpretation can be generalized to more complicated systems.

wave mechanics   A quantum-mechanical theory introduced by Schrödinger in 1926 which ascribes wave characteristics to the fundamental entities of atomic structure, and formulates the appropriate wave equation (Schrödinger's equation).

wave noise   Noise in the current of a detector, caused by fluctuations in the electromagnetic radiation falling on the detector (cf. shot noise).

wavenumber   The reciprocal of the wavelength; i.e., the number of waves per unit distance in the direction of propagation.

wave zone (also called far field)   The field of a pulsar beyond the velocity-of-light radius.

weak-equivalence principle   A principle derived from the equality of the inertial and gravitational mass, which states that if we observe two bodies experiencing equal accelerations, we cannot, by observing the motion, tell whether they are being subjected to a uniform acceleration by some external mechanism or whether they happen to be in a uniform gravitaitonal field.

weak-field condition (gravitational) $\phi << c^2$, where ϕ is the Newtonian gravitational potential.

weak interaction   A short-range nuclear force responsible for radioactivity and for the decay of certain unstable nuclei, e.g., $e^- + p \rightleftharpoons n + \nu_e$, which is so called because it occurs at a rate slower than that of the strong interaction by a factor of about 10^{-13} (see interactions).

weber (Wb)   The derived SI unit of magnetic flux. 1 weber = 10^8 maxwells.

Werner lines   Spectral lines of molecular hydrogen in the ultraviolet, in the same general region as the Lyman lines.

Wesselink analysis   A method of determining the radius of a variable star.

Whirlpool galaxy   A spiral galaxy (M51, NGC 5194) of type Sc in Canes Venatici.

whistlers   Radio waves generated by a flash of lightning, which travel along Earth's magnetic field out beyond the ionosphere and back to Earth. They arrive back with a descending pitch because the high-frequency end of the wave train arrives first (see dispersion).

white dwarf (wd or D)   A star of high surface temperature, low luminosity, and high density (10^5–10^8 g cm^{-3}), with roughly the mass of the Sun and the radius of the Earth, that has exhausted most or all of its nuclear fuel, believed to be a star near its final stage of evolution. When the Sun becomes a white dwarf, its radius will be about 0.01 of its present radius. DA white dwarfs are hydrogen-rich; DB white dwarfs are helium-rich; DC are carbon rich; DF are calcium-rich; DP are magnetic stars. White dwarfs have relatively low rotational velocities.

Widmanstätten pattern   A geometric pattern found in some iron meteorites, consisting of groups of parallel lamellae crossing each other at various angles.

Wien's law   The wavelength at which a blackbody emits the greatest amount of radiation is inversely proportional to its absolute temperature.

Wilson-Bappu effect   A linear relation between the width of the K_2 emission core in the resonance line of Ca II at 3933 Å, detectable for late-type stars, and the absolute magnitudes of the stars.

window   A term used to describe the spectral range within which the Earth's atmosphere is transparent to radiation (see optical window and radio window). Earth's atmosphere is completely opaque to X-rays; ultraviolet radiation is absorbed by electronic transitions in the ozone layer, but it is possible to get above some of it in balloons and rockets. Infrared radiation is absorbed by water vapor, so a high mountain or desert will let some radiation through (the vibrations of molecules cause absorption in the near-infrared, and the rotation of molecules causes absorption in the far-infrared and short radio). Wavelengths beyond the radio window are absorbed by the ionosphere.

WKB method (Wentzel-Kramers-Brillouin)   A method for obtaining an approximate solution to Schrödinger's equation.

Wolf 359   A nearby flare star.

Wolf diagram   Logarithmic plot of N (number of stars or galaxies counted at successive magnitude limits) versus apparent magnitude.

Wolf-Lundmark system   A dwarf E5 elliptical galaxy, sometimes considered a member of the Local Group.

Wolf number (R)   (Also called relative number.) A quantity which gives the number of sunspots, and the number of groups of sunspots, at a given time. $R = k(10g + f)$, where k is a constant depending on observing conditions, g is the number of sunspot groups, and f is the number of individual spots visible on the Sun at a given time.

Wolf-Rayet (W-R) star   One of a class of very luminous, very hot (as high as 50,000 K) stars whose spectra have broad emission lines (mainly He I and He II), which are presumed to originate from material ejected from the star at very high ($\sim$2000 km $^{-1}$) velocities. Some W-R spectra show emission lines due to carbon (WC stars); others show emission lines due to nitrogen (WN stars). (Hiltner and Schild classification: WN-A, narrow lines; WN-B, broad lines.)

Wollaston prism   A prism used to obtain plane-polarized light.

work function (W)   The amount of energy needed to release an electron from the attraction of positive ions in a metal. It is different for different metals.

world line   The graph in spacetime coordinates which represents any continuous sequence of events relating to a given particle. In general relativity, all material particles have timelike world lines, photons have null world lines, and tachyons have spacelike world lines.

world model   A mathematical model of the Universe.

world point   See event.

X

X-band   A radio band at a wavelength of 3.7 cm (8085 MHz).

X-ogen   An unidentified molecular transition at 3.36 mm (89.19 GHz) discovered in 1970.

X-rays   Photons of wavelengths between about 0.1 Å and 100 Å—more energetic than ultraviolet, but less energetic than γ-rays.

X-ray astronomy   A new field of astronomy which studies (from rockets and satellites) the wavelength region from about 0.1 to 50 Å. This region cannot be studied from Earth's surface because our atmosphere is very opaque to radiation of these wavelengths.

X-ray pulsars   Pulsars (q.v.) that radiate in the X-ray region of the spectrum. Best verified examples are Her X-1 and Cen X-3. They are thought to be rotating, strongly magnetic neutron stars of about 1 $M\odot$ in a grazing orbit around a more massive star from which they are accreting matter.

X-ray sources   A class of celestial objects whose dominant mechanism of energy dissipation is through X-ray emission. Galactic X-ray sources appear optically as starlike objects, peculiar in their ultraviolet intensity, variability (on time scales ranging from milliseconds to weeks), and spectral features. All known compact X-ray sources are members of close binary systems; a current popular model is mass accretion onto a compact object from a massive companion. (Four X-ray sources—all variable—are known to be associated with globular clusters.) The 21 known extended X-ray sources associated with clusters of galaxies seem to be clouds of hot gas trapped in the cluster's gravitational field.

Y

Yerkes system   See MKK system.
young disk Cepheids   Type I Cepheids.

Z

z   See redshift.
Z-number   See atomic number.
z pinch   A diffuse toroidal pinch in which the magnetic field runs around the plasma column.
Zanstra's theory   A theory of emission lines in planetary nebulae which supposes that the emission lines in hydrogen (and helium) arise from a process of ionization (by the ultraviolet radiation of the central star) and recombination, and that the forbidden lines arise from the collisional excitations of the metastable state.
Zeeman effect   Line broadening due to the influence of magnetic fields. A multiplet of lines is produced, with distinct polarization characteristics. The Zeeman effect is measured by measuring the difference between right-hand and left-hand polarization across a spectral line.
zenith   The point on the celestial sphere directly above the observer's head—i.e., opposite to the direction of gravity (cf. nadir).
Zerilli's equation   A Schrödinger-type equation for even-parity perturbations on the Schwarzschild metric.
zero-age main sequence   The position on the H-R diagram for stars which have attained hydrostatic equilibrium and have started hydrogen burning in their cores, but which have not yet had time to produce an observable change in their chemical composition.
zero-point energy   The energy of the lowest state of a quantum system. Amount of vibrational energy allowed by quantum mechanics to be associated with atomic particles at 0 K, whereas classical mechanics requires this to be zero. Also, the energy of an electron in its ground state.
zero-point pressure   The pressure contributed by degenerate electrons, which do not come to rest even at absolute zero.
zeroth law of thermodynamics   See thermal equilibrium, law of.

zodiac   A band about 8° wide on the celestial sphere, centered on the ecliptic.

zodiacal light   A faint glow that extends away from the Sun in the ecliptic plane of the sky, visible to the naked eye in the western sky shortly after sunset or in the eastern sky shortly before sunrise. Its spectrum indicates it to be sunlight scattered by interplanetary dust. (*Pioneer 10* has determined that its brightness varies inversely as the square of the distance out to 2.25 AU and then decreases more rapidly.) The zodiacal light contributes about a third of the total light in the sky on a moonless night.

zone of avoidance   An irregular zone near the plane of the Milky Way where the absorption due to interstellar dust is so great that no external galaxies can be seen through it.